T0181466

Studies in Computational Intelligence

Volume 553

Series editor

Janusz Kacprzyk, Polish Academy of Sciences, Warsaw, Poland
e-mail: kacprzyk@ibspan.waw.pl

For further volumes:
http://www.springer.com/series/7092

About this Series

The series "Studies in Computational Intelligence" (SCI) publishes new developments and advances in the various areas of computational intelligence—quickly and with a high quality. The intent is to cover the theory, applications, and design methods of computational intelligence, as embedded in the fields of engineering, computer science, physics and life sciences, as well as the methodologies behind them. The series contains monographs, lecture notes and edited volumes in computational intelligence spanning the areas of neural networks, connectionist systems, genetic algorithms, evolutionary computation, artificial intelligence, cellular automata, self-organizing systems, soft computing, fuzzy systems, and hybrid intelligent systems. Of particular value to both the contributors and the readership are the short publication timeframe and the world-wide distribution, which enable both wide and rapid dissemination of research output.

Roger Y. Lee

Editor

Applied Computing and Information Technology

 Springer

Editor
Roger Y. Lee
Software Engineering and Information
   Technology Institute
Central Michigan University
Mt. Pleasant, MI
USA

ISSN 1860-949X   ISSN 1860-9503   (electronic)
ISBN 978-3-319-34403-4   ISBN 978-3-319-05717-0   (eBook)
DOI 10.1007/978-3-319-05717-0
Springer Cham Heidelberg New York Dordrecht London

Printed on acid-free paper

Springer is part of Springer Science+Business Media (www.springer.com)

Preface

The purpose of the First International Symposium on Applied Computing and Information Technology (ACIT 2013) held on August 31–September 4, 2013 in Matsue City, Japan, was to bring together researchers, scientists, engineers, industry practitioners, and students to discuss, encourage, and exchange new ideas, research results, and experiences on all aspects of Applied Computers and Information Technology, and to discuss the practical challenges encountered along the way and the solutions adopted to solve them. The conference organizers have selected the best 12 papers from those accepted for presentation at the conference in order to publish them in this volume. The papers were chosen based on review scores submitted by members of the program committee and underwent further rigorous rounds of review.

In "Intelligent Billboard Based on Ambient System (IBBAS)", Baker Alrubaiey, Morshed Chowdhury, and Atul Sajjanhar have designed and developed an Intelligent Bill Board based on Ambient System. This board is capable of interacting with humans in close proximity. A number of sensor devices are used in the board where sensors inputs (Bluetooth IDs and distance sensor readings) create an innovative form of user interaction with the board.

In "Deriving Pauses for Obtaining Fundamental Movements in Traditional Skills", Teruhisa Hochin and Hiroki Nomiya try to derive pauses in order to obtain fundamental movements, which are considered to be important for transmitting traditional skills, from the movement of a worker or a player. This is due to the observation that the fundamental movements may relate to pauses. The fundamental movements are captured as similar changes in the speeds appearing everywhere.

In "A Novel Approach to Design of an Under-Actuated Mechanism for Grasping in Agriculture Application", Alireza Ahrary and R. Dennis A. Ludena present a novel approach to mechanical design of underactuated robot finger with passive adaptive grasping. Three concepts of environment, mechanism, and control of passive adaptive grasping are also focused in our approach. The mechanical design of proposed underactuated robot finger is briefly described and experimental results in real environments are also given.

In "Discovering Unpredictably Related Words from Logs of Scholarly Repositories for Grouping Similar Queries", Takehiro Shiraishi, Toshihiro Aoyama, Kazutsuna Yamaji, Takao Namiki, and Daisuke Ikeda present a method to find related query words at the first step from logs of scholarly repositories. In particular, we try to find words which are related from the viewpoint of non-researchers. In this sense, these words are unpredictably related. A simple method to do this using the access log is that we treat queries which lead to the same paper as related.

In "A Trichotomic Approach to Concept Capture and Representation: With its Application to Library Data Mining", Toshiro Minami, Sachio Hirokawa, Kensuke Baba, and Eriko Amano propose a method for specifying the concept that is too hard to describe in an exact way by a word or a phrase, by setting up the "relative distances" from three key concepts which we call a trichotomic approach to concept capture and representation, or description, by approximate means. Secondly, they demonstrate the usefulness of the trichotomic approach through a couple of case studies applied to library's loan record analysis.

In "Hard Optimization Problems in Learning Tree Contraction Patterns", Yasuhiro Okamoto and Takayoshi Shoudai discuss two optimization versions of the MINL problem, which are called MINL with Tree-size Maximization (MAX MINL) and MINL with Variable-size Minimization (MIN–MAX MINL). We show that MAX MINL is NP-complete and MIN–MAX MINL is MAX SNP-hard.

In "A Web Page Segmentation Approach Using Seam Degree and Content Similarity", Jun Zeng, Brendan Flanagan, Qingyu Xiong, Junhao Wen, and Sachio Hirokawa propose two parameters: seam degree and content similarity, to indicate the coherent degree of a page block. Instead of analyzing predefined heuristics or visual cues, our method utilizes the visual and content features to determine whether a page block should be divided into smaller blocks. We also proposed a principled page segmentation method using these two parameters.

In "Improving Particle Swarm Optimization Algorithm and Its Application to Physical Travelling Salesman Problems with a Dynamic Search Space", Benoît Vallade and Tomoharu Nakashima address an improvement idea for the Particle Swarm Optimization Algorithm (PSO) and its implementation on a Traveller Salesman Problem-based competition. As a search algorithm, the PSO is used to tune a set of parameters, which usually take their values in static search spaces. This chapter proposes a solution to use effectively the PSO algorithm on optimization problems using parameters that take their values in dynamic space.

In "Experimental Implementation of a M2M System Controlled by a Wiki Network", Takashi Yamanoue, Kentaro Oda, and Koichi Shimozono discuss an Experimental implementation of an M2M system, which is controlled by a wiki network. This M2M system consists of mobile terminals at remote places and wiki servers on the Internet. A mobile terminal of the system consists of an Android terminal and it may have an Arduino board with sensors and actuators. The mobile terminal can read data from not only the sensors in the Arduino board but also wiki pages of the wiki servers.

In "Personal Ontology Extraction Considering Content Concordance from Tagging to Webpages in Similar SBM Users", Fumiko Harada and Hiromitsu Shimakawa have studied a method to extract hierarchical and synonymous relationships among tagged phrases on a social bookmark (SBM) for an individual

SBM user. It detects the relationships from webpage clusters with the same tagged phrases derived from the bookmarks shared in the target and his similar SBM users. However, noisy tagging violating personal phrase meaning degrades its detection accuracy. This chapter proposes a method to improve such drawback.

In "Psychophysiological and Behavioral Evaluation of the Process of Mastering Skills: To Select Appropriate Indices for a Target Movement", Hiroko Sawai, Kazune Tomotake, Yasuharu Ishii, Keisuke Ueno, and Emi Koyama present a study to evaluate the early process of mastering skills to reveal skill factors including pause and adjustment in producing traditional handicrafts using psychophysiological and behavioral indices. The indices were measured in two experiments. Tasks that needed obtaining skill factors were performed. As a result, heart rates, time-series behaviors of electrooculogram (EOG), and wrist activities during tasks, and subjective scores before tasks reflected the difference in the product quality in the early process of mastering skills.

In "Frameworks for Adaptive Human Management Systems Based on MDA", Haeng-Kon Kim and Roger Y. Lee describe the initial investigation in the fields of MDA and generative approaches to SOA. Our view is that MDA aims at providing a precise framework for generative software production. Unfortunately many notions are still loosely defined (PIM, PSM, etc.). We propose here an initial exploration of some basic artifacts of the MDA space to SOA. Because all these artifacts may be considered as assets for the organization where the MDA is being deployed with SOA, we are going to talk about MDA and SOA abstract components to apply an e-business application.

It is our sincere hope that this volume provides stimulation and inspiration, and that it will be used as a foundation for works to come.

Satoshi Takahashi

Contents

Contributors

Alireza Ahrary Faculty of Computer and Information Sciences, Sojo University, Kumamoto, Japan

Baker Alrubaiey School of Information Technology, Deakin University, Melbourne Campus, Australia

Eriko Amano Office for eResource Services, Kyushu University Library, Fukuoka, Japan

Toshihiro Aoyama Department Electronic and Information Engineering, Suzuka National College of Technology, Mie, Japan

Kensuke Baba Research and Development Division, Kyushu University Library, Fukuoka, Japan

Morshed Chowdhury School of Information Technology, Deakin University, Melbourne Campus, Australia

Brendan Flanagan Graduate School of Information Science and Electrical Engineering, Kyushu University, Fukuoka, Japan

Fumiko Harada Department of Information Science and Engineering, Ritsumeikan University, Kusatsu, Shiga, Japan

Sachio Hirokawa Research Institute for Information Technology, Kyushu University, Fukuoka, Japan

Teruhisa Hochin Department of Information Science, Kyoto Institute of Technology, Kyoto, Japan

Daisuke Ikeda Department of Informatics, Kyushu University, Fukuoka, Japan

Yasuharu Ishii Graduate School of Science and Technology, Kyoto Institute of Technology, Kyoto, Japan

Haeng-Kon Kim Department of Computer Engineering, Catholic University of Daegu, Gyeongsan-si, Korea

Emi Koyama Graduate School of Science and Technology, Kyoto Institute of Technology, Kyoto, Japan

Roger Y. Lee Software Engineering and Information Technology Institute, Central Michigan University, Mt. Pleasant, USA

R. Dennis A. Ludena Department of Computer and Information Sciences, Sojo University, Kumamoto, Japan

Toshiro Minami Kyushu Institute of Information Sciences, Fukuoka, Japan

Tomoharu Nakashima Department of Computer Science and Intelligent Systems, Osaka Prefecture University, Osaka, Japan

Takao Namiki Department of Mathematics, Hokkaido University, Hokkaido, Japan

Hiroki Nomiya Department of Information Science, Kyoto Institute of Technology, Kyoto, Japan

Kentaro Oda Computing and Communications Center, Kagoshima University, Kagoshima, Japan

Yasuhiro Okamoto Department of Informatics, Kyushu University, Fukuoka, Japan

Atul Sajjanhar School of Information Technology, Deakin University, Melbourne Campus, Australia

Hiroko Sawai Graduate School of Science and Technology, Kyoto Institute of Technology, Kyoto, Japan

Hiromitsu Shimakawa Department of Information Science and Engineering, Ritsumeikan University, Kusatsu, Shiga, Japan

Koichi Shimozono Computing and Communications Center, Kagoshima University, Kagoshima, Japan

Takehiro Shiraishi Department of Informatics, Kyushu University, Fukuoka, Japan

Takayoshi Shoudai Department of Informatics, Kyushu University, Fukuoka, Japan

Kazune Tomotake Graduate School of Science and Technology, Kyoto Institute of Technology, Kyoto, Japan

Keisuke Ueno Graduate School of Science and Technology, Kyoto Institute of Technology, Kyoto, Japan

Benoît Vallade Department of Computer Science and Intelligent Systems, Osaka Prefecture University, Osaka, Japan

Junhao Wen Graduate School of Software Engineering, Chongqing University, Chongqing, China

Qingyu Xiong Graduate School of Software Engineering, Chongqing University, Chongqing, China

Kazutsuna Yamaji National Institute of Informatics, Tokyo, Japan

Takashi Yamanoue Computing and Communications Center, Kagoshima University, Kagoshima, Japan

Jun Zeng Graduate School of Software Engineering, Chongqing University, Chongqing, China

Intelligent Billboard Based on Ambient System (IBBAS)

Baker Alrubaiey, Morshed Chowdhury and Atul Sajjanhar

Abstract An Intelligent Bill Board Based on Ambient System (IBBAS) has been designed and developed. This board is capable of interacting with humans in close proximity. A number of sensor devices are used in the board where sensors inputs (Bluetooth IDs and distance sensor readings) create an innovative form of user interaction with the board. The IBBAS display is determined by user position, location, and movements. In this chapter, the authors investigate how the user inputs are mapped to the advertising board and its behavior. A prototype of IBBAS is implemented.

Keywords Ubiquitous computing · Billboard · Application · Bluetooth · Phidget

1 Introduction

Ambient intelligence is a new concept that allows interaction between humans and computers through the usage of ubiquitous computing devices. Ambient Intelligence shares ideas with Ubiquitous Computing, Ubiquitous Communication and Intelligent User Interface [1]. Ubiquitous computing are defined as many computers serving one person. Ubiquitous Communication is defined as wireless networks technology used for communication in Ambient Intelligent Systems. For Intelligent User Interfaces, sensors can be used to detect motion, voice, etc. instead of using a mouse or keyboard to communicate with the system [1].

B. Alrubaiey · M. Chowdhury (✉) · A. Sajjanhar
School of Information Technology, Deakin University, Melbourne Campus, Australia
e-mail: muc@deakin.edu.au

B. Alrubaiey
e-mail: balrubai@deakin.edu.au

A. Sajjanhar
e-mail: atuls@deakin.edu.au

R. Y. Lee (ed.), *Applied Computing and Information Technology*,
Studies in Computational Intelligence 553, DOI: 10.1007/978-3-319-05717-0_1,
© Springer International Publishing Switzerland 2014

The IBBAS is developed as a prototype application that is capable of interpreting sensor based human interaction to control a computer application driving what is displayed on a board. The study investigates a novel combination of sensor inputs (Bluetooth IDs and distance sensor readings) to create an innovative form of user interaction with the board. Representations of different combinations of user inputs (position, location, and user movements) are developed, and how the inputs map influences board behavior is determined.

The study also explores how to identify a user profile using the system and how to display appropriate advertisements according to the user's interest and interactions. Major guidelines for installing the board in public and private places are provided. The study also covers the positioning of displays, the brevity of glances, content type, what catches the eye, content format and dynamics, and the relative merits of small displays versus large displays are discussed.

The remainder of this chapter is organized as follows: The Literature Review (Sect. 2) sets out different work related to interactive advertising boards. The study covers MirrorBoard, MagicBroker, Speakeasy, and Bluetooth Scanning. The Bluescreen System, GroupCast, MagicBoard, FunkyWall, the Interactive Public Ambient Displays and the major guidelines for installing the board in public and private places are all discussed. The concept and design of IBBAS (Sect. 3) looks at Technologies Used such as BlueCove and Phidgets, the events used when the pre-condition fulfilling the events fires the triggers, the architecture of the IBBAS and the flow chart of the system. The IBBAS Implementation and experimentation (Sect. 4) discusses the Sequence Diagram and Algorithms are given to describe IBBAS with UML design diagrams, BlueCove API and IBBAS Applications, Phidget API and IBBAS Application and Experimentation. The Conclusion (Sect. 5) summarizes the whole chapter and (Sect. 6) sets out future development.

2 Related Work

Schönböck et al. [2] describe MirrorBoard as a system that "grabs people's attention" via the board and allows users to interact with the board though mechanisms such as switching between advertisements and placing the user's pictures into the screen. A disadvantage of MirrorBoard is that some people may not like to have their picture on a public display. User cannot interact with the board effectively. Also the system is very slow in changing the picture on display because the camera used at that time has frame rate limited to 15 frames per second.

Erbad et al. [3] developed a large public interactive screen called MagicBroker, which is a large screen that can interact with the user via his/her mobile phone. MirrorBoard is more advanced than MagicBroker and MagicBoard because it has multimodal interactions with a user. The MirrorBoard has a camera to capture the image of a person and can interact explicitly with the user by using Bluetooth or implicitly by using a microphone. MagicBroker and MagicBoard have similarities in that both implement SMS messages from a mobile phone or PDA and board.

A disadvantage of MagicBroker is that the system cannot be used by everyone because it is relies only on mobile device interactions. Therefore, the system can be limiting for someone without a mobile device. The system is not very efficient in advertising because passerby may not be aware about the system and not be aware of how they might interact with the system.

Izadi et al. [4] developed a public display called Speakeasy that connects a public display with any media. Speakeasy is a platform which provides data as sources and receives data known as a sink. A disadvantage of using Speakeasy with the LED sign is that it only allows text-based content and does not support graphics. Graphical Display System used wireless keyboard-mouse unit for interaction with soup to add or retrieve data which is not allow the user to interact from long distance without keyboard or mouse.

Schmidt et al. [5] describe a technique called Bluetooth scanning which detects Bluetooth IDs when users (with Bluetooth phones enabled, for example) are close to the board and displays advertisements of personal interest for the specific user(s). Speakeasy allows users to send to the board and can display on the screen. Bluetooth scanning can give more information about a user and the content of the display which MirrorBoard cannot. A disadvantage of using Bluetooth scanning is lack of reliability because not everyone has a mobile phone or Bluetooth when passing the board. Voice recognition may be not very reliable and so the system may not recognize the user and display the required advertisement. People dislike being tracked and recognized in public places by using their picture or voice.

Sharifi et al. [6] developed a BluScreen that detects Bluetooth IDs for any device such as a mobile phone or PDAs and displays appropriate advertisements based on the ID. A disadvantage of BluScreen is that user profiles are not implemented in the system so we couldn't improve the content of display. Also, the system doesn't know the activity patterns for each user, and of course, a user without a Bluetooth device cannot interact with the board at all.

GroupCast [7] uses infrared badges to interact with the public displays for a nearby audience. GroupCast can be used for conversations between groups of individuals and displaying appropriate content to the users according to their profiles in common areas of interest. GroupCast can create user profiles for identifying the user and Bluetooth scanning advances active badges for users to interact with the board. A disadvantage of using GroupCast is that it takes a long time to match user profiles with all other users and to find the intersection of user preferences.

FunkyWall [8] is an interactive board; the user can interact with it by gesture and speech. FunkyWall uses more than one technique for interaction with a board such as gesture, sound (speech) and visuals but there are difficulties for users when interacting with the board. A disadvantage of using FunkyWall is that a participant has to wear sensor gloves to be able to interact with the board. User cannot interact with the system from long distance.

Vogel et al. [9] designed an interactive public ambient display system that supports transition from implicit to explicit interaction with both public and personal information. A user can interact with the board by using hand gestures and touch screen inputs which are used for explicit interaction; body orientation and position cues are used for implicit interaction. Interactive Public Ambient Display is more advanced than FunkyWall because it has implicit and explicit detection for a user approaching the board. A disadvantage is that it is very difficult for the user to control the board by different hand movements, showing private information on public displays such as e-mail and personnel details (which may not be comfortable for some users) and the system needs explicit interaction such as touching screen for more complex interaction.

MyAdvertisements [10] is an advertisement technology that can be used in a building equipped with sensorial and computational capabilities such as Wi-Fi connections, RFID tags and the user should carry portable devices such as PDAs or Smartphone to be able to interact with the system. MyAdvertisements is not suitable for use in every building because it requires wireless technologies such as Wi-Fi or RFID tags equipped for a building. Also, user information needs to be stored in the system, and then the system is able to display the required advertisements. This limitation of the MyAdvertisements system makes it unsuitable to be used in public display area without any Wi-Fi or RFID devices.

There are general guidelines for installing interactive digital advertising boards either in public or private places such as the position of the board, the size of board, and the contents of the display which should all be taken into consideration. Huang et al. [11] suggest the location and content of the board should be taken into account during the design of the board. The content of the display should not take more than a few seconds for the user to engage with as the content of the display attracts passersby within the surrounding environment. There are many factors affecting the attraction of people to an advertising board. These are the simplicity of use of the technology, the positioning of displays, the brevity of glances, content type, content format and dynamics and the use and merits of small displays versus large displays [11].

The combination of eight Phidget distance sensors and Bluetooth has not been explored for interactive ad boards, and this chapter explores this novel combination. We also explored what types of gestures and motions can be effectively captured using this configuration as well as how they could be combined with limited personalization via Bluetooth IDs. The actual physical configuration and arrangement of the sensors near the board is investigated and an algorithm is proposed for the user inputs action.

3 Architecture and Design of IBBAS

This section outlines the information and concepts related to developing the IBBAS. It sets out the technologies used (Sect. 3.1) such as Bluetooth and Phidget sensors, the events (Sect. 3.2) used to notify the applications that can be used to fire the condition when the pre-condition is fulfilled, the architecture of IBBAS (Sect. 3.3) is explored and a flow chart (Sect. 3.4) is introduced to explain how the system detects the user.

Fig. 1 Event-fire conditions

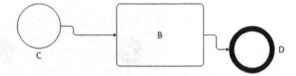

3.1 Hardware Used

The following technologies were used in developing IBBAS. Technologies include Bluetooth, Phidget sensors and the Java language for implementing the IBBAS application prototype.

3.1.1 Bluetooth

Bluetooth wireless technology is an open specification that is publicly available and can communicate using radio waves at a distance of 10 m. Also a Bluetooth device supports the data and voice for communication and operates at 2.4 GHz so it can be used anywhere because it has a free license industrial, scientific, and medical (ISM) band.

3.1.2 Phidget Sensors

A Phidget sensor is a device that has the ability to measure some physical quantity by converting an analogue signal measure to a digital signal. Also, a sensor can be used to sense the physical world and uses sensor data as input data for computation processing. The sensors are classified according to their sensitivity such as light, motion, temperature, magnetic fields, gravity, humidity and other physical aspects of the external environment. A Phidget Interface Kit (PIK) [12] is used for the IBBAS prototype implementation. The PIK consists of 8 analogue inputs which are the IR Distance Sensor which is attached through an analogue port on PIK, and used to measure continuous sensor readings of the distance of an object from the sensor from 10 to 70 cm. The Phidget interface Kit is connected to a PC through a USB connection.

3.2 Event-Driven

Event handlers are used to notify the application that an event has occurred. Fire condition is a concept used for sensor interpretation which is essentially a single rule that can be compared to sensor input to give a true or false value. Figure 1 illustrates the fire condition concept, the passive elements (C) are called "condition" and the active elements are a "B event". The arrow from the condition "C" to the event "B"

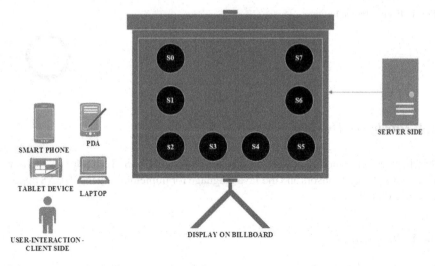

Fig. 2 Architecture of client-serve IBBAS system

indicate that "C" is a precondition for event "B". Also an arrow from the event "B" to the condition "D" shows that "D" is the post-condition to the event "B". Every condition is either fulfilled (indicated by a token—black dot inside the circle) or unfulfilled at a given point in time. The number of tokens is not conserved in the firing of an event. Therefore a single pre-condition can rise to any number of post-conditions or many pre-conditions may finally cause just a single post-condition to be true. If the pre-condition that fulfilled the event B is activated it can occur (or fire) but when the post-condition is fulfilled the event is not activated.

3.3 Architecture of IBBAS

Figure 2 illustrates the architecture of the IBBAS application which consists of a client side (which is just a mobile device with Bluetooth capabilities) and the server side that comprises the advertisements board with eight Phidget sensors which have the ability to detect the user motion and a computer connected to the board which runs the IBBAS application. Figure 3 illustrates the hardware tools used in constructing the architecture of the IBBAS.

3.3.1 DataFlow Diagram

Figure 4 illustrates the flow chart diagram of the system, starting with the idle state when no Bluetooth ID detected. When a Bluetooth ID detects a new device, the system checks whether the device has registered with the system or not. If not, the system goes back to an idle state. If yes, the user can interact with the Phidget sensor and this triggers the display of the ads. While the user is interacting with the board,

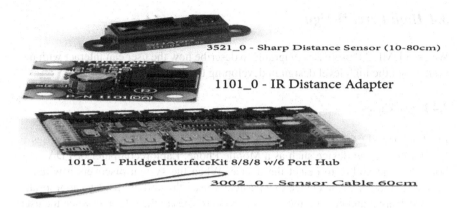

3521_0 - Sharp Distance Sensor (10-80cm)

1101_0 - IR Distance Adapter

1019_1 - PhidgetInterfaceKit 8/8/8 w/6 Port Hub

3002_0 - Sensor Cable 60cm

Fig. 3 Phidget hardware for IBBAS system

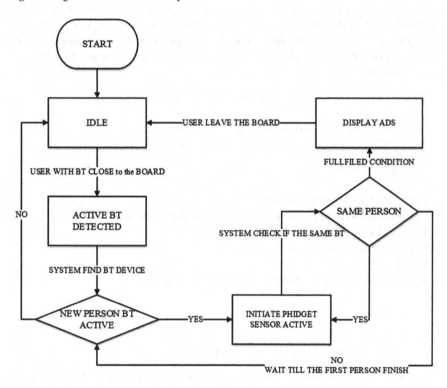

Fig. 4 Flow chart of the IBBAS system

the system can identify if the same person is near the board or another new device for a different user is near the board. The system waits for the current person to finish, and then a new person can interact with the system. When the user leaves the board, the system goes back to the idle state, if no other user is present.

3.4 High Level Design

We used UML 2.0 use-cases diagrams to describe how the user can interact with the system and the high level design in developing the IBBAS application.

3.4.1 Use Cases

In Use-Case (1) the user can interact with the system when the user carries a Bluetooth-enabled device such as a Mobile Phone, Laptop computer or PDA. The board has the ability to detect the device. When the system discovers a wireless device, and matches the device to a particular user, the system displays appropriate ads which are matched with the user. The system goes to the idle state when the user either leaves the board or switches the device off. When the system discovers a new device which is not registered with the system, the system will not interact with the user (and could notify user to enter the device id into the system to be able to interact appropriately). The following illustrates the use-case diagram which assumes that the device id has been entered into the system and the user is close enough to interact with the board.

1. We assume the user device id has been entered into the system

   Use Case Name: Discover device close to Ads Board.
   Trigger/Goal: To allow user to interact with Ads board.
   Actors: User
   Main Flow:

1. User switches his/her device on.
2. System detects device and check device Id for particular user.
3. User interact with the ads boards.
4. System display required ads for user.
5. User left the Board.
6. System switches off ads and puts the board in the idle state.

   Extensions:

   2a. Device is new:
      2a1. System notifies User to enter Device Id.

In Use-Case (2) operator can enter into the system new user to be able to interact with the system. The operator inter the details of new user such as device-id, name and phone number, and the system save the details and gives permission to the user to interact with system

2. We assume the user is not registered in the system

   Use Case Name: Add a new Customer
   Trigger/Goal: To enter the details of a new Customer

Actors: Operator
Main Flow:

1. Operator enters customer Device ID.
2. System validates that ID is new.
3. Operator enters name and phone number.
4. System saves the details. *Extensions*:

   2a. ID is not new:
      2a1. System notifies Operator, and terminates the use case.

4 Implementation and Experimentation

This section outlines the implementation of the IBBAS application prototype and the experimentation used in developing the IBBAS system.

4.1 Algorithm of the IBBAS Application and Sequence Diagram

Pseudo code for the IBBAS prototype implementation

1. Create an event listener that will respond to a Bluetooth inquiry related event.
2. Search for a Bluetooth device and if found return the device discovered.
3. Create Phidget object.
4. Use a function to handle the event.
5. Pass this function to the event dispatcher.
6. Activate the Phidget and wait for events on the distance sensors.
7. Interpret sensor readings as triggers.
8. Map triggers to selected action.
9. Perform the action (display a pre- defined PowerPoint slide representing an advertisement).
10. System goes to idle state when no one is close to the board, as detected by no devices discovered from the continual Bluetooth inquiries.

Figure 5 is a sequence diagram which commences when a user switches Bluetooth on, and then the device is discovered by the Bluecove DiscoveryListener. When the listener finds a device it informs the inquiry CompleteEvent and the event object initiates the Phidget sensors to interact with the user. When the sensor reading inputs match the conditions, the SensorChangeEvent will inform the event object that these conditions match and then the event will fire the condition and display the relevant PowerPoint slide. The p: Process performs the display action from the operating system of the window.

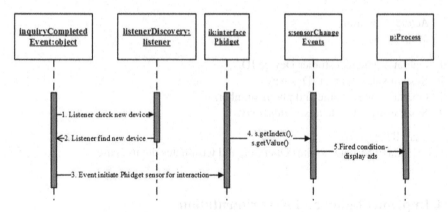

Fig. 5 IBBAS sequence diagram

4.2 BlueCove API and IBBAS Application

Bluecove is a Java library for Bluetooth [13] (Java Specification Request 82 (JSR-82)) which can interface with many operating systems such as window XP SP2. The design outlined in Sect. 3 shows the important protocols addressed by Java APIs for Bluetooth wireless technology (JABWT) such as Service Discovery Protocol (SDP), which is used in an IBBAS application prototype. Bluetooth is used for wireless connection. Therefore, a way to find devices to connect to and a way to learn what those devices can do was required. The API provides a way to discover devices, find services and undertake services registration.

- Interface javax.bluetooth.DiscoveryListener

This class provides an event listener that will respond to inquiry-related events and service search. deviceDiscovered( ) is called each time a device is found.

- class javax.bluetooth.DiscoveryAgent

This class provides methods for service and device discovery. startInquiry( ) method to place the local device in inquiry mode, retrieveDevices( ) method to return information about devices and cancelInquiry( ) method to cancel an inquiry.

Either one or two packages are used for implementation of the IBBAS application for Connected Limited Device Configuration (CLDC). The core package is javax.bluetooth and the second package contains javax.obex which depends on the javax.microedition.io.

4.3 Phidget API and IBBAS Application

Events were discussed because Phidget, the API library [12], employs events extensively. Figure 6 illustrates the flowchart diagram for creating an event and uses that

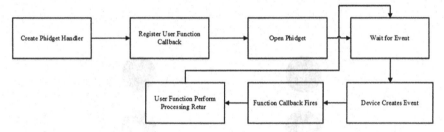

Fig. 6 Flow chart for phidget event

event to be fired. In the IBBAS application prototype, the first step is to discover the Bluetooth ID by the deviceDiscovered( ) method. This can be used as an event, then the event handler is created and that event is registered with the Phidget API library. The library can call an event handler function whenever the event occurs. Subsequently the Open( ) function will signal the Phidget library when the Phidget is attached. A sensorChanged( ) method is used to set the sensitivity of sensors between 250 and 450 so the sensor reading input should be lesser than 450 and higher than 250. When the sensor reading inputs match that condition the trigger will be fired and advertisements will be displayed with a third party application, PowerPoint.

While the user is standing near the board and moving in front or across the sensors, the action would be false and the display will not change. If the user moves backward or forward so that the distance from the board is less or more than the sensor inputs required the board will display nothing.

We chose sensor-reading inputs of 250–450 that are the sensitivity level of the sensor reading according to the user distance from the sensor. While the user stands close to the board the user moves continuously forward or backward. Therefore I have chosen the range of sensor reading inputs between 250 and 450 which allow triggers to be fired and advertisements to be displayed. The sensor reading is 450 when the user stands approximately 0.3 m from the board but the sensor reading changes to 250 when the user is closer than 0.3 m from the board. In other words the sensor reading increases when the user is closer to the board but sensor readings decrease when the distance between the user and the board is increased.

4.4 Experimentation: Exploring Interaction Possibilities with IBBAS

The architecture of IBBAS has four zones where each can interact with the board, these are:

1. zone (1): sensor (0), sensor (1) and sensor (2) on the left hand side of the board,
2. zone (2): sensor (2) and sensor (3) on the middle bottom of the LHS of the board,
3. zone (3): sensor (4) and sensor (5) on the middle of the RHS of the board, and
4. zone (4): sensor (5), sensor (6) and sensor (7) on the right hand side of the board.

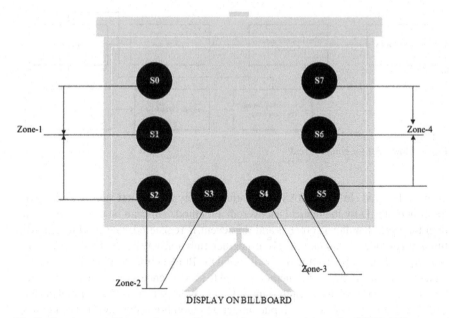

Fig. 7 User interact with *zone-1* or *zone-2* or *zone-3* or *zone-4* and the action DISPLAYS

The sensor reading input measures the distance of the user from the board and uses these measurements as triggers to perform the required actions. Users can interact with the board by standing close to the LHS of the board with a distance of 0.3 m from the board; the user should have a mobile phone with Bluetooth capabilities enabled. When this precondition is satisfied, the trigger fires and the system display the required advertisements. Figure 7 illustrates how users interact within zone (1) or zone (2) or zone (3) or zone (4) and the action following the fired condition which is to open a PowerPoint slide to display required ads.

4.5 Class Diagram

Figure 8 illustrates the IBBAS class diagram. The IBBAS prototype implementation initiated by the method deviceDiscovered( ) is called each time a device is found during an inquiry. Also, this interface allows an application to specify an event listener that will respond to inquiry-related events and used for service searching. When the inquiry is completed or canceled, this method is called INQUIRY_COMPLETED, INQUIRY_ERROR or INQUIRY_TERMINATED constant to differentiate between completed, error and canceled inquiries. The servicesDiscovered( ) method allows the user to interact with the Phidget sensor. This method creates Phidget object which is driven by an event handler created by the deviceDiscovered( ) method which activates the Phidget and waits for events.

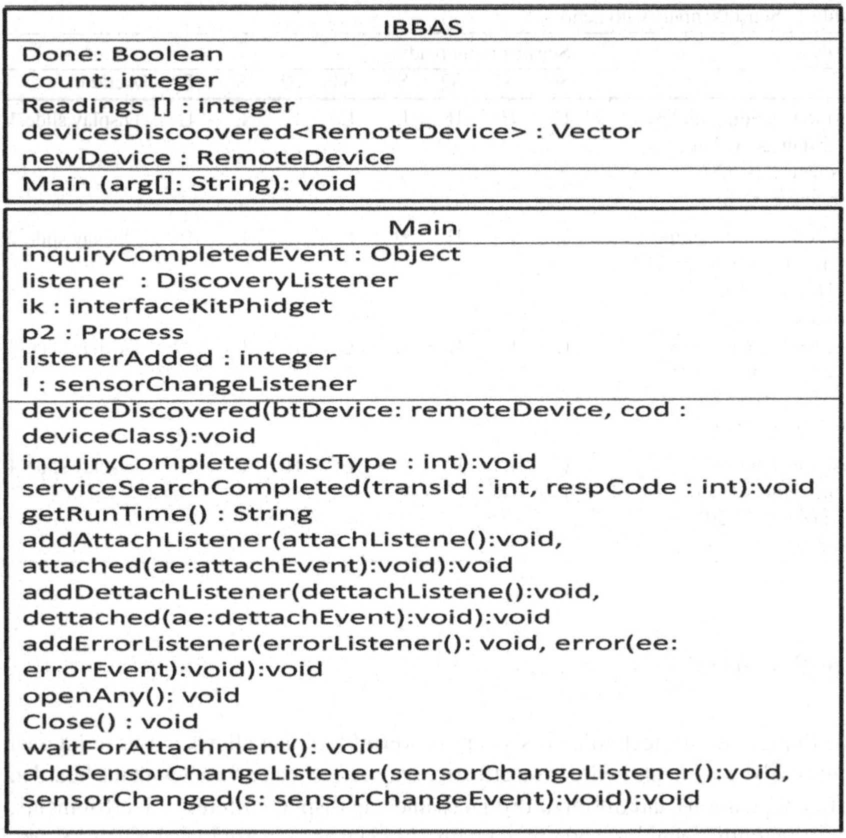

IBBAS
Done: Boolean
Count: integer
Readings [] : integer
devicesDiscoovered<RemoteDevice> : Vector
newDevice : RemoteDevice
Main (arg[]: String): void

Main
inquiryCompletedEvent : Object
listener : DiscoveryListener
ik : interfaceKitPhidget
p2 : Process
listenerAdded : integer
I : sensorChangeListener
deviceDiscovered(btDevice: remoteDevice, cod : deviceClass):void
inquiryCompleted(discType : int):void
serviceSearchCompleted(transId : int, respCode : int):void
getRunTime() : String
addAttachListener(attachListene():void, attached(ae:attachEvent):void):void
addDettachListener(dettachListene():void, dettached(ae:dettachEvent):void):void
addErrorListener(errorListener(): void, error(ee: errorEvent):void):void
openAny(): void
Close() : void
waitForAttachment(): void
addSensorChangeListener(sensorChangeListener():void, sensorChanged(s: sensorChangeEvent):void):void

Fig. 8 IBBAS class diagram

• Algorithm of the IBBAS execution

Pseudo code for the IBBAS prototype implementation

1. Create an event listener that will respond to Bluetooth inquiry related event.
2. Search for Bluetooth device and if found return the device discovered.
3. Create Phidget object.
4. Use a function to handle your event.
5. Pass this function to the event dispatcher.
6. Activate the Phidget and wait for events on the distance sensors.
7. Interpret sensor readings as triggers.
8. Map triggers to selected action.
9. Perform the action (display a pre-defined power point slide representing an advertisement).
10. System goes to idle state when no one close to the board, as detected by no devices discovered from the continual Bluetooth inquiries.

Table 1 Sensors inputs with actions

Trigger	Sensors input reading								
	S0	S1	S2	S3	S4	S5	S6	S7	Action
Position-standing up left-distance—0.3 m H (range 250–520) L (range 0–150)	H	H	H	L	L	L	L	L	Display slide-1
Position-position-sitting down left-bottom—0.3 m H (range 250–520) L (range 0–150)	L	L	H	H	L	L	L	L	Display slide-2
Position-sitting down right-bottom—0.3 m H (range 250–520) L (range 0–150))	L	L	L	L	H	H	L	L	Display slide-3
Position-standing up right—0.3 m H (range 250–520) L (range 0–150)	L	L	L	L	L	H	H	H	Display slide-4

4.6 Results

The Phidget sensor technology is a very powerful tool that allows a user to interact with an advertisements board in an effective way. A user can interact with the board in different positions and different distances and can display different advertisements. The Bluetooth technology combined with Phidget sensors provides the IBBAS prototype application with flexibility regarding where, when and how the advertisements should be displayed on the board. The Phidget distance sensors have very high sensitivity that can detect any unwanted moving objects such as trees and animals and it gives various input readings. Therefore, I took sensor reading inputs (any showing a reading of sensitivity of over 50 from the sensor) to ignore unwanted readings, and to allow the user to interact without interruption from any other object movements. Overall the IBBAS system allows a user to interact with the board smoothly and display the required advertisements in any format such as a PowerPoint presentation or any other application.

Table 1 shows the sensors input reading and the corresponding action when the condition is satisfied. The first inputs reading are S0 = High, S1 = High, S2 = High, S3 = Low, S4 = Low, S5 = Low, S6 = Low and S7 = Low, the action displays PowerPoint slide-1. Sensor reading inputs can be high or low depending on the condition of the user input on the sensors. The second trigger is sensors (s2 , s3) when the user changes not distance but he/she changes position; instead of standing, the user can be in a sitting position and when the condition is satisfied, action-2 is performed which is to display PowerPoint slide-2. The third trigger is sensors (s4, s5) when the user changes location to the RHS of the board and the user can be in a sitting position.

When the condition is fired the action-3 is performed which is to display PowerPoint slide-3. The fourth trigger occurs when the user moves from left hand side to the right hand side; he/she will interact with sensors (s5, s6, s7) so those sensors will have high readings and the rest will have low related readings. Therefore, a different action will be performed for the different trigger. In addition, we changed the distance of the user from the board from 0.3 to 0.5 m and performed similar experiments to produce another four triggers.

For example the user can interact within zone (2) when the user changes his/ her position such as sitting down or in a position to interact with the sensors of the middle bottom of the LHS of the board (S2, S3). Figure 9 illustrates how users interact with the zone (2) sensors and the action following the fired condition which is to open a power point slide to display a different corresponding slide. Ads can then be chosen by the user by positioning seat accordingly as we discuss below.

5 Conclusion

The literature review considered related work undertaken on interactive boards, including their architecture, design and implementation attempted. IBBAS was developed based on ideas from previous works. This research paper outlines the concept related to developing an IBBAS such as events used to notify the application. Also the chapter describes the process of developing the IBBAS prototype based on the technologies such as Bluetooth detection and Phidget sensor interpretation. IBBAS has the ability to detect the presence of humans close to the board as well as sensing Bluetooth IDs of the user. The implementation of the prototype used Java language, Phidget API and Bluecove API. Phidget sensors are a set of sensor devises that detect the distances of the user from the board combined with Bluetooth sensor capabilities to make the program effective in capturing a range of user gestures and movements. Bluetooth IDs are considered for creating an event to initiate Phidget sensors, i.e. the sensors are activated by the presence of a user (detected via Bluetooth device discovery), assuming the user carries a Bluetooth enabled device. The combination of eight Phidget distance sensors and Bluetooth has not been explored for interactive ad boards, and this project explored this novel combination. I also explored the types of gestures and motions that can be effectively captured using this configuration as well as how it could be combined with limited personalization via Bluetooth IDs. The actual physical configuration and arrangement of the sensors near the board was developed, as well as a suitable algorithm to process the user inputs action. The experiment section explored how the application works and how the user can interact with the system. The experimentation provided us with inputs for sensors to illustrate a number of interaction possibilities with IBBAS.

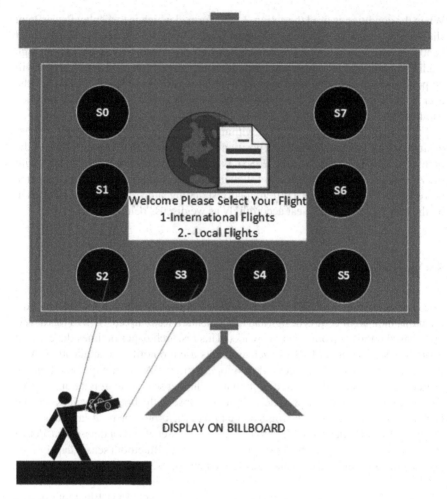

Fig. 9 User interact with zone-2 and the action DISPLAYS

6 Future Work

The aim of the IBBAS project was to create an application using a Phidget sensor and Bluetooth ID that allows users to interact with advertisements on the billboard. This application has limitations that allow only one user to interact with the billboard at a time but future work may enable more than one user at a time. Only eight sensors were used in developing IBBAS but future work may use more. The Phidget Interface Kit allows more than eight sensors through the use of a USB hub. Bluetooth ID was used for activating Phidget sensors to allow the user to interact with the Phidget sensors but future work may use the Bluetooth ID for more purposes such as sending SMS or personalization of advertisements to the system. The Bluetooth ID can be used to store advertisements for particular users and displayed when the user interacts with the board.

References

1. G. Riva, F. Vatalaro, F. David, M. Alcaniz, *Ambient Intelligence* (IOS Press, Amsterdam, 2005), pp. 5–6
2. J. Schönböck, F. König, G. Kotsis, D. Gruber, E. Zaim, A. Schmidt, MirrorBoard—an interactive billboard, in *Mensch and Computer*, 2008, pp. 217–226
3. A. Erbad, M. Blackstock, A. Friday, R. Lea, J. Al-Muhtadi, MAGIC broker: a middleware toolkit for interactive public displays, in *PerCom*, 2008, pp. 509–514
4. E.W.S. Izadi, J.Z. Sedivy, T. Smith, J. Black, J.I. Hong, M.W. Newman, in *Speakeasy: A Platform for Interactive Public Displays*, 2008
5. A. Schmidt, F. Alt, P. Holleis, J. Müller, A. Krüger, Creating log files and click streams for advertisements in physical space, in *Adjunct Proceedings of Ubicomp*, Seoul, Korea, 2008, pp. 18–19
6. M. Sharifi, T. Payne, E. David, Public display advertising based on bluetooth device presence, in *MIRW*, 2006
7. J.F. McCarthy, T.J. Costa, E.S. Liongosari, Unicast, outcast and groupcast: three steps toward ubiquitous peripheral displays, in *Proceedings of 3rd International Conference on Ubiquitous Computing (Ubicomp)*, UK, 2001, pp. 332–345
8. A. Lucero, D. Aliakseyeu, J.B. Martens. Funky wall: presenting mood boards using gesture, speech and visuals, in *ACM AVI'08*, Napoli, Italy, pp. 425–428
9. D. Vogel, R. Balakrishnan, Interactive public ambient displays: transitioning from implicit to explicit, public to personal, interaction with multiple users, in *UIST*, 2004, pp. 137–146
10. A. Di Ferdinando, A. Rosi, R. Lent, A. Manzalini, F. Zambonelli, MyAdvertisements: a system for adaptive pervasive advertisements, in *Pervasive and Mobile computing*, 2009, pp. 385–401
11. E.M. Huang, A. Koster, J. Borchers, Overcoming assumptions and uncovering practices: when does the public really look at public displays?, in *Pervasive*, 2008, pp. 228–243
12. Documents, http://www.phidgets.com/docs/Main_Page. Accessed 30 April 2012
13. Remote Device Discovery, http://www.bluecove.org/bluecove/apidocs/overview-summary. html%23DeviceDiscovery. Accessed 25 April 2012

Deriving Pauses for Obtaining Fundamental Movements in Traditional Skills

Teruhisa Hochin and Hiroki Nomiya

Abstract This chapter tries to derive pauses in order to obtain fundamental movements, which are considered to be important for transmitting traditional skills, from the movement of a worker or a player. This is due to the observation that the fundamental movements may relate to pauses. The fundamental movements are captured as similar changes of the speeds appearing everywhere. The movement is represented with a sequence of the sum of the speeds of the parts of a body in this chapter. Subsequences are cut out from the whole sequence. The similarity of two subsequences is judged in the frequency domain. It is experimentally shown that pauses can easily be obtained by adjusting several parameters of the similarity test. The typical fundamental movement is tried to be obtained.

Keywords Pause · Traditional skill · Fundamental movement · Frequency

1 Introduction

There are many traditional crafts and industries in Japan. In almost all of them, things are made by hand, without the use of machinery. The craft workers joining traditional industries have special techniques and skills. It takes a long time to obtain them. It is, however, becoming difficult for the next generation to inherit them because young people seldom join the traditional crafts and industries [1, 2]. Analyzing the techniques and skills may clarify their important points. These points could make the learning time shorter than usual. This will make it easy and quick for young people interested in the traditional crafts and industries to learn such techniques and skills.

T. Hochin (✉) · H. Nomiya
Department of Information Science, Kyoto Institute of Technology, Kyoto, Japan
e-mail: hochin@kit.ac.jp

H. Nomiya
e-mail: nomiya@kit.ac.jp

R. Y. Lee (ed.), *Applied Computing and Information Technology*,
Studies in Computational Intelligence 553, DOI: 10.1007/978-3-319-05717-0_2,
© Springer International Publishing Switzerland 2014

In order to transmitting traditional techniques and skills from the current generation to the next one, several efforts have been made. These include the analysis of the movement of craft work [1–3], the construction of skill training system based on data mining and visualization [4], that of the prototype self-training system using a three dimensional depth sensor and a camera [5], that of an interactive multimodal motion learning system [6], and that of the augmented practice mirror learning support system [7].

Recently, the pause, which is called "Ma" in Japanese, has been focused on in order to derive the skills from the master's movement, and to transmit them to the next generation. The movements of a craft worker have been analyzed from the point of view of the pause [3]. A method of representing the pause has been proposed [8].

Here, Nakamura has intensively studied the pause especially in speeches and music pieces [9]. Nakamura defined the pause. She emphasized that a pause is not merely a silent period, but gives some impression to an audience. She revealed the characteristics of the pause mainly from the points of view of the impression and the aspect of physiological psychology [9].

Various pauses have been clarified in various performances [10]. Pauses in speeches, music pieces, and actions have been obtained. We have categorized the pauses into several types from the various points of view: the presence of sound, that of movement, that of a pose, the time duration, and the aim of a pause. It has also been pointed out that there is a negative pause, whose time duration is shorter than the usual one. It is shown that short pauses are considered to be important for transmitting traditional skills.

This chapter tries to derive pauses for obtaining fundamental movements often appearing in the movements of traditional workers. They are derived by using the frequency characteristics. It is experimentally shown that they can be derived by choosing several parameters of the similarity test.

The remaining of this chapter is as follows: Sect. 2 describes the related works on the pauses and the frequency analysis. Section 3 describes the tea ceremony, whose movements we treat as the traditional work in this chapter. Section 4 experimentally derives pauses. Some considerations are made in Sect. 5. Finally, Sect. 6 concludes the chapter.

2 Related Works

2.1 Pauses

Pauses in speeches and music pieces. The pause means no movement. It is the temporary stop. Nakamura revealed and emphasized its importance [9]. Nakamura defined the appropriate pause of a speech and/or a music piece as the time required and sufficient for understanding the contents and the meaning of the phrases and/or notes preceding it, for connecting to the next phrases and/or notes, and for representing some impression and receiving it.

Nakamura also revealed the relationships between the pause and the breath [9]. It has experimentally been revealed that the length of a pause is based on that of a breath. Synchronizations of the breaths of a music player and an audience, and those of a speaker and a listener are shown.

Pauses in traditional skills. Tanaka et al. have studied on the pauses in the knitting of wire netting [3]. The purposes of their study were to find the pauses in the knitting process of wire netting, and to reveal their effect. A knitting work is composed of the main tasks and the secondary ones. The main task is the movement of the knitting. The other movements are included in the secondary ones. The main tasks alternate with the secondary ones.

Pauses were tried to be found when fingers stop in knitting wire netting. Fingers, however, did not stop in the task. They defined a pause as the duration when the acceleration of the motion of a finger is in some range. According to this definition, pauses could be found in around 15–40 % of the time after a main task started.

They concluded that the pauses appeared in the working time, and that the pauses were related to short breaks of finger's motion. These pauses are important for skilled workers to repeat the same movement for a long time.

Kinds of pause. Pauses have been categorized into several types from the following points of view: the time duration, the presence of sound, that of movement, that of a pose, and the aim of a pause [10].

Time duration. Pauses are divided into three categories in the point of view of the length: short, normal, and long pauses. The duration of a normal pause is about 1 s, which corresponds to one breath. That of a short (long, respectively) pause is shorter (longer) than 1 s. Short pauses appear in work, e.g. the knitting of wire netting. On the other hand, long ones appear in performances, e.g. speeches, music pieces, and Kabuki performances.

Sound. Pauses are categorized into two types according to the accompaniment of sound: a pause without sound, and a pause with sound. The former is an ordinary pause. A pause of a speech is an example of this type of pause. An example of the latter type of pause is the note with a fermata.

Movement. In actions, there are two types of pause: a pause without movement, and a pause with movement. The former means there is no movement. An actor stops his/her motion. This type of pause is an ordinary pause. On the other hand, the pauses in the knitting of wire netting are of the latter type of pause because fingers move.

Poses. There are two types of pause on the existence of a pose: a pause with a pose, and a pause without a pose. The former is a usual pause in an action. An actor stops his/her motion. The latter type of pause has two meanings. One includes neither movements nor poses. An actor does not do anything. The other includes movement. The pauses of the knitting of wire netting are of this type of pause.

Aim. Pauses could further be divided to two types according to the aims of pauses: a pause for an audience, and a pause not for an audience. The former is for giving some impression to someone. A player gives an audience some impression. The pauses in speeches and music pieces are of this kind of pause. The latter is not for anyone. The pause of the knitting of wire netting is of this type of pause. A skilled person does not intend to give any impression to anyone.

2.2 Deriving Patterns from Time-Series Data

For the purpose of deriving patterns from time-series data, similarity of two sequences of data must be judged. There are two major approaches in judging the similarity of the sequences. One compares two sequences directly in the time domain [6, 11–13]. The Adaptive Piece-wise Constant Approximation (APCA) [11] approximates a sequence into a sequence of steps, and uses the areas between the steps and the axis. The Multi-resolution Vector Quantized (MVQ) [12] divides a sequence into segments, and approximates it by using a sequence of the representative values of the segments. The ratio of the representative values of a candidate sequence to those of a key one is their similarity. These methods, however, could not properly handle the sequence slightly different from the one in the time direction. For example, the similarity of a sequence and the slightly shifted one becomes very low. In order to adapt this situation, the dynamic programming approach is often taken. The Dynamic Time Warp (DTW) e.g. [13] is a popular method. Ozaki and Oka followed this approach to extract knacks of skills [6].

The other approach compares two sequences in the frequency domain [14–16]. The sequence of data is transformed to a set of data in the frequency domain. The Fourier transformation is often used as such a transformation. It has been shown that similar sequences can be retrieved by using a few Fourier coefficients [14]. This also brings the dimensionality reduction. We can construct an index for fast retrieval by using only these coefficients [15, 16].

By applying Fourier transformation to a sequence of data, Fourier coefficients are obtained. These except for the 0th one are complex numbers, which is composed of a real number and an imaginary one, while the 0th coefficient is a real number. The 0th coefficient represents the mean value of the data in the sequence. By using these coefficients, we could compare two segments in the frequency domain. The coefficients of low orders correspond to those of low frequencies, while those of high orders correspond to those of high frequencies. By using coefficients of several lowest orders (all of orders, respectively), we could approximately (precisely) compare two segments.

In general, we could define and use the distance between two segments. The distance is the dissimilarity of two segments. We can use the distance in judging whether they are similar, or not. We could decide two segments are similar if the distance is less than a value, which is usually called the threshold value. It is the boundary of deciding whether two segments are similar, or not.

When the time series is very long, we often use the window to obtain their subsequences. The Fourier transformation is applied to a subsequence in the window. The subsequence is called a segment. In the window approach, segments are obtained by shifting the window. The amount of the shift of the window Ls as well as the width of the window Lw should be considered. When Ls is larger than or equal to Lw, there is no overlap between two windows. There are some overlaps among windows when Ls is smaller than Lw.

3 Tea Ceremony as Traditional Works

The movements of tea ceremony are used as those of traditional works. After the summary of tea ceremony is presented, the pauses in tea ceremony are described.

3.1 Tea Ceremony

In tea ceremony, a host serves cups of tea for guests. There are strict rules on how to prepare, make, and serve tea. Tea ceremony consists of many steps.

Scenes of tea ceremony are shown in Fig. 1. These are of the preparation phase of making a cup of tea. A host makes the instruments used in the tea ceremony clean by using Fukusa, which is a sheet of cloth. The host begins to prepare (Fig. 1a). He takes the Hishaku, the ladle for scooping water, up (Fig. 1b), and puts it down (Fig. 1c). He bows to the guest (Fig. 1d). He puts a tea container down in front of him (Fig. 1e). He opens Fukusa (Fig. 1f) to fold it. This action is called the Fukusa-sabaki. The tea container is purified (Fig. 1g). He puts it down in front of a water jug (Fig. 1h). He opens Fukusa to begin Fukusa-sabaki again (Fig. 1i).

3.2 Pauses in Tea Ceremony

There are two kinds of pause. Motions entirely stop in one kind of pause, while there are some movements in the other kind of pause.

In the former kind of pause, a host intends to make pauses, and to divide the series of actions. A host may take a breath at this kind of pause. The scenes of these pauses are those shown in (Fig. 1a, b, d, f). A kind of impression is received at these scenes. The length of the pause is from a half to one second. This pause corresponds to the normal pause.

On the other hand, a host does not intend to make pauses in the latter kind of pause. These are considered to be stays rather than pauses. These pauses do not divide the series of actions. The scenes shown in (Fig. 1c, e, g–i) are of this type of pause. A host may not take breaths at these pauses. These are caused by that a host politely treats the instruments. These pauses are shorter than the former ones. This pause is considered to be the short pause.

The sums of the speeds of the twenty-nine parts of the body are shown in Fig. 2 in the normalized form. There are many periods where speeds are equal to zero. These periods are considered to be of both types of pause. These types, however, could

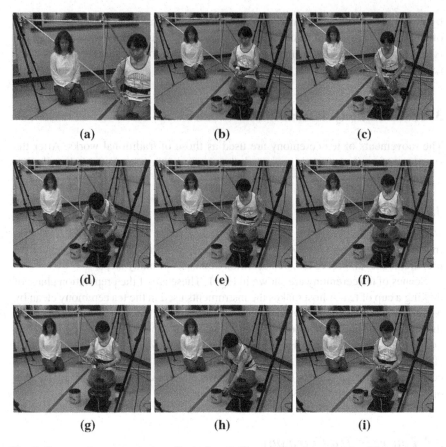

Fig. 1 Scenes of tea ceremony: **a** Beginning, **b** Kagami-Bishaku, **c** Putting Hishaku down, **d** Bowing, **e** Putting a tea container down in front of a host, **f** Fukusa-sabaki, **g** Purifying a tea container, **h** Putting a tea container down in front of a water jug, and **i** Opening Fukusa again

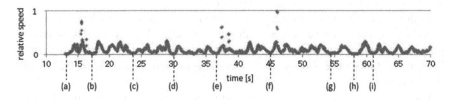

Fig. 2 Normalized speeds of movements of tea ceremony

not be distinguished each other only by considering the length of a pause. This is because the time duration of the longest short pause is very close to that of the shortest normal one [10]. It has been shown that the scenes of the normal pauses could be distinguished from the short pauses by using the fact that the host straightens himself or bows in the normal pauses [10]. Here, both types of pause are tried to be derived.

4 Derivation of Fundamental Movements

Fundamental movements of tea ceremony are tried to be obtained by focusing on pauses. Here, we follow the approach comparing sequences in the frequency domain. The Fast Fourier Transformation (FFT) is applied to the time series of speeds. As the whole of the movements of tea ceremony is very long, the window method is used.

4.1 Experimental Methods

The sampling frequency of the three-dimensional motion captured data used in the experiments is 100 Hz. The duration of a time slice is 0.01 s.

We conduct three experiments. The first is to decide the appropriate width of the window Lw. The second is to decide the number of Fourier coefficients used in calculating the similarities of segments. The last is to determine the threshold value in judging the similarity of two segments.

The segment of the movement (e) of Figs. 1 and 2 is used as a key segment of retrieval because this is a typical one of the pauses not for an audience in tea ceremony. The segments similar to the key segments are retrieved.

In the last two experiments, the goodness of the selection of the value of the parameter is evaluated by using the measures of precision, recall, and F-measure. These are calculated as

precision $= Ncr/Nr,$
recall $= Ncr/Nc,$ and
F-measure $= 2*$precision$*$recall$/($precision $+$ recall$),$

where Ncr is the number of the correct segments retrieved, Nr is the number of the retrieved segments, and Nc is the number of the correct segments. The correct segments are manually decided by one of the authors. In the experiments, the number of the correct segments is sixteen because there are fifteen segments similar to the key segment, which is the segment of the movement (e) of Fig. 2.

Experiment 1. As described in Sect. 3, it is considered that fundamental movements exist in short and/or normal pauses. The duration of a normal pause is about 1 s. The candidate widths of the window Lw are 32, 64, 128, and 256. Here, we use the numbers of powers of two because the number of points must be a power of two in using FFT.

We adopt the width of the best retrieval as Lw. In this experiment, we set Ls be 10.

Experiment 2. It is experimentally decided the number of Fourier coefficients used in calculating the similarity of segments. The number is varied from two to four. Adopting two coefficients means that the 0th and the 1st coefficients are used. We use 128 as Lw because it gives the best performance as described later. The goodness of the number of Fourier coefficients used is evaluated through precision, recall, and F-measure.

Table 1 Precision, recall, and F-measure against the number of Fourier coefficients used

Number of Fourier coefficients	Precision	Recall	F-measure
2	1.00	0.81	0.90
3	1.00	0.94	0.97
4	0.89	1.00	0.94

Table 2 Precision, recall, and F-measure against threshold values

Threshold value	Precision	Recall	F-measure
1,000	1.00	0.63	0.77
1,500	1.00	0.94	0.97
2,000	0.88	0.94	0.91

Experiment 3. We experimentally decide the threshold value. We try 1,000, 1,500, and 2,000 as the threshold values. Three Fourier coefficients are used because they attain the best as described later. The goodness of the threshold value is evaluated by using precision, recall, and F-measure.

4.2 Experimental Results

Experiment 1. The results are shown in Fig. 3. The sequences of bold red squares show the candidates obtained by the retrieval. In the cases of 32 and 64, many sequences do not become the candidates similar to the key sequence. These lengths seem to be too short to capture the waveform of the key sequence. In the case of 256, the length of the segment is too long to represent the waveform of the key segment. On the other hand, the similar waveforms are obtained in the case of 128. It is considered that 128 is better than 32, 64, and 256.

Experiment 2. The results are shown in Table 1. In the case of two and three coefficients, several subsequences do not become the candidates similar to the key sequence. In the case of four coefficients, the sequences not so similar to the key sequence are obtained. Totally, the best performance could be obtained when the three coefficients are used. We decided to use three coefficients.

Experiment 3. The results are shown in Table 2. In the cases of 1,000 and 1,500, several sequences do not become the candidates similar to the key sequence. In the case of 2,000, the sequences not so similar to the key sequence are obtained. Totally, the best performance could be obtained when the threshold value is 1,500. The value 1,500 is adopted as the threshold value.

The segments of bold red squares shown in Fig. 4 are those obtained when the window size Lw is 128, three coefficients are used, and the threshold value is 1,500.

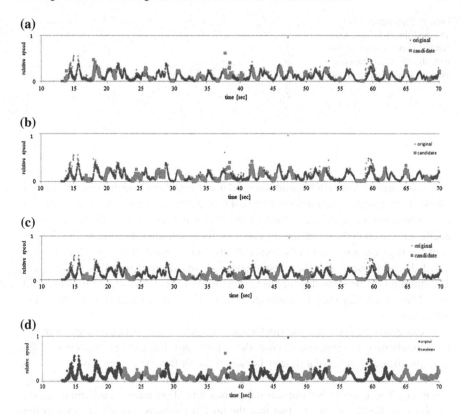

Fig. 3 Movements similar to the one of the scene **e** shown in Figs. 1 and 2. Window widths *Lw* are **a** 32, **b** 64, **c** 128, and **d** 256

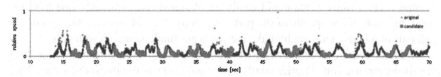

Fig. 4 Segments obtained by the retrieval when the window size *Lw* is 128, three coefficients are used, and the threshold value is 1,500

5 Considerations

When we use 128 as the window size *Lw*, the best performance could be obtained. As the length of a normal pause is about one second, and the sampling frequency is 100 Hz, this value is considered to be appropriate. When four Fourier coefficients were used in calculating the distance between two segments, the precision degraded. As the third coefficient is a high frequency component, it is considered that the effect of the high frequency component is not good in calculating the distance between segments.

Fig. 5 The waveform
obtained through the inverse
Fourier transformation only
with the three coefficients
used in comparing segments

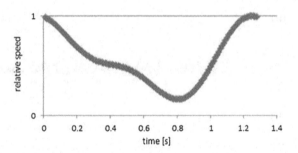

As the three Fourier coefficients are used in comparing segments, we could obtain
the waveform used in the comparison. That is, the waveform used in the comparison
can be obtained through the inverse Fourier transformation only with the three Fourier
coefficients used in the comparison. The waveform obtained is shown in Fig. 5. This
is of the key segment used in the experiments, the segment of the movement (e) in
Fig. 2. It can be seen that the relative speed slowly decreases and then increases.
The duration of the slow-down phase seems to be longer than that of the speed-up
one. This movement may be considered to be the fundamental movement of tea
ceremony.

It can easily be seen that the fundamental movements appear everywhere in the
movements of tea ceremony. Even if the actions are different, the patterns of the
movements are quite similar. It may be one of the skills in tea ceremony.

The method described in this chapter could derive fundamental movements of tea
ceremony. These are in both of normal pauses and short ones. The normal pauses
can be identified by using the fact that the host straightens himself or bows in the
normal pauses [10]. Therefore, we could distinguish the fundamental movements of
the normal pauses from those of the short ones.

Fundamental movements could easily be obtained. We treat only one kind of
speed, the sum of the speeds of the parts of a body. We did not treat the speeds of
each part of a body. This might make it easy to get fundamental movements.

We decided to use three Fourier coefficients and 1,500 as the threshold value. We
may, however, use four Fourier coefficients and 2,000 as the threshold value. The
values within some range may be acceptable.

6 Conclusion

This chapter tried to derive pauses in order to obtain fundamental movements from the
movement of a worker or a player. The movement is represented with a sequence of
the sum of the speeds of the parts of a body. The similarity of two subsequences cut out
from the whole sequence was judged in the frequency domain. It was experimentally
shown that pauses could easily be obtained by adjusting the width of the window, the
number of Fourier coefficients used, and the threshold value in judging the similarity
of two segments. The typical fundamental movement in tea ceremony was tried to
be obtained.

Only the part of the preparation phase of tea ceremony was treated in this chapter. Treating the whole of tea ceremony is in future work. The movements treated in this chapter are represented with sequences of the sums of the speeds. The movements of each part of a body are not considered. A part of a body resides at a point in three-dimensional space. We may have to treat the movements as those in the spatio-temporal space. Spatio-temporal analysis of the movement is also in future work. Tea ceremony is treated in this chapter. Trying other traditional works is also included in future work.

Acknowledgments This research is partially supported by the Ministry of Education, Science, Sports and Culture, Grant-in-Aid for Scientific Research (B), 23300037, 2011–2014.

References

1. K. Morimoto, N. Kuwahara, Holistic analysis on affective source of Japanese traditional skills in Japan, in *Proceedings of 1st International Conference on Affective and Pleasurable Design (APD2012)* (2012), pp. 5239–5243
2. M. Kume, T. Yoshida, Characteristics of technique or skill in traditional craft workers in Japan, in *Proceedings of APD2012* (2012), pp. 5289–5297
3. T. Tanaka, A. Ohnishi, M. Shirato, M. Kume, K. Tsuji, A. Goto, A. Nakai, T. Yoshida, Interval of weaving kanaami, in *Dynamics and Design Conference* (2008), pp. 326_1–326_5 (in Japanese)
4. M. Ozaki, N. Oka, A training system for skill learning based on mining and visualization of skill archives, vol. 105(639), *IEICE Technical report* (2006), pp. 5–10 (in Japanese)
5. N. Kuwahara, K. Morimoto, J. Ota, M. Kanai, J. Maeda, M. Nakamura, Y. Kitajima, K. Aida, Sensor system for skill evaluation of technicians, in *Proceedings of APD2012* (2012), pp. 5254–5263
6. M. Araki, Multimodal motion learning system for traditional arts, in *Proceedings of APD2012* (2012), pp. 5274–5281
7. I. Kuramoto, Y. Inagaki, Y. Shibuya, Y. Tsujino, *Augmented Practice Mirror: A Self-learning Support System of Physical Motion with Real-Time Comparison to Teacher's Model, LNCS*, vol. 5620 (Springer, Berlin, 2009), pp. 123–131
8. T. Hochin, Y. Ohira, H. Nomiya, Representation and management of physical movements of technicians in graph-based data model, in *Proceedings of APD2012* (2012), pp. 5264–5273
9. T. Nakamura, Psychological study of 'ma' (a synonym of 'pause') in communication. J. Phonetic Soc. Japan **13**, 40–52 (2009). (in Japanese)
10. T. Hochin, H. Nomiya, Analysis of puases toward transmitting traditional skills, in *Proceedings of 14th ACIS International Conference on Software Engineering, Artificial Intelligence, Networking and Parallel/Distributed Computing (SNPD2013)* (2013), pp. 414–419
11. E.J. Keogh, K. Chakrabarti, S. Mehrotra, M.J. Pazzani, Locally adaptive dimensionality reduction for indexing large time series databases, in *Proceedings of 2001 ACM SIGMOD International Conference on Management of Data* (2001), pp. 151–162
12. V. Megalooikonomou, Q. Wang, G. Li, C. Faloutsos, Multiresolution symbolic representation of time series, in *Proceedings of 21st IEEE International Conference on Data Engineering* (2005), pp. 668–679
13. S.-W. Kim, S. Park, W.W. Chu, An index-based approach for similarity search supporting time warping in large sequence databases, in *Proceedings of 17th IEEE International Conference on Data Engineering* (2001), pp. 607–614
14. R. Agrawal, C. Faloutsos, A.N. Swami, Efficient similarity search in sequence databases, in *Proceedings of the 4th International Conference on Foundations of Data Organization and Algorithms (FODO'93)* (1993), pp. 69–84

15. C. Faloutsos, M. Ranganathan, Y. Manolopoulos, Fast subsequence matching in time-series databases, in *Proceeings of 1994 ACM SIGMOD International Conference on Management of Data* (1994), pp. 419–429
16. T. Hochin, Y. Yamauchi, H. Nomiya, H. Nakanishi, M. Kojima, Fast subsequence matching in plasma waveform databases, in *Proceedings of fifth International Conference on Intelligent Information Hiding and Multimedia, Signal Processing (IIH-MSP2009)* (2009), pp. 759–762

A Novel Approach to Design of an Under-Actuated Mechanism for Grasping in Agriculture Application

Alireza Ahrary and R. Dennis A. Ludena

Abstract Under-actuated mechanism is a mechanical system which the number of control inputs is smaller than the Degrees of Freedom (DOF). This mechanism has many advantages such as low cost, low energy consumption and lightweight. This chapter present a novel approach to mechanical design of under-actuated robot finger with passive adaptive grasping. Three concepts of environment, mechanism and control of passive adaptive grasping are also focused in our approach. The mechanical design of proposed under-actuated robot finger is briefly described and experimental results in real environments are also given.

Keywords Under-actuated · Mechanical design · Robot finger

1 Introduction

Challenges of developing robotic hands, which are capable of grasping a wide variety of objects, are simple control structure with a high number of DOF and self-adaptive functionality without reducing the performance. In most application areas, manipulation of objects with complex hands is often not essential and grasping devices are sufficient. However, these kinds of hands are not capable of adapting of the shape of different objects in most cases. A possible approach to simple control structure is that of reducing the number of actuators getting more efficient, simpler and reliable than their fully actuated alternatives [1]. But the dexterity of the grasp is affecting by reducing the number of actuators and dimensions of the force and motion. Some robot hands like Robonaut Hand [2] and DLR Hand II [3, 4] are developed

A. Ahrary (✉)
Faculty of Computer and Information Sciences, Sojo University, Kumamoto, Japan
e-mail: ahrary@ieee.org

R. D. A. Ludena
Department of Computer and Information Sciences, Sojo University, Kumamoto, Japan
e-mail: dennis@cis.sojo-u.ac.jp

R. Y. Lee (ed.), *Applied Computing and Information Technology*,
Studies in Computational Intelligence 553, DOI: 10.1007/978-3-319-05717-0_3,
© Springer International Publishing Switzerland 2014

with the aim of imitating the dexterity and adaptability of robot hand to improve the grasping functions. These dexterous robot hands are big, heavy and not suitable for using in real life application.

Recently, neuroscience studies on human hands inspired new research on control strategies and design for robotic hands. These studies are focus in to achieve a trade-off between simplicity, control of DOF and its versatility [5, 6]. The synergy idea has been also applied to control different hand models such as simple gripper, the Barrett hand, the DLR hand, the Robonaut hand and the human hand model [6]. Brown et al. proposed a design of robot hand that is able to match postural synergies mechanically coupling motion of the single joints [5]. Wimboeck et al. introduce a synergy impedance controller, which is derived and implemented on the DLR Hand II [7]. A robot hand with many DOF can exploit postural synergies to control force and motion of the grasped object is investigated in [8]. Evaluation of grasping quality can be defined by different performance measures [9], which are classified in three groups. Considering the properties of the grasp matrix and the position of the contact points [10]. Considering the manipulation configuration [11] and taking into account both the kinetic properties of the grasped object and of the manipulators [9].

This chapter presents a new under-actuated robot finger with passive adaptive grasping. In design of these robot finger we take the advantage of under-actuation, i.e. the number of actuators are less than active joints, and combined with novel under patent mechanism for self-adaptive functionality.

2 Previous Works

In the work relevant to the automatic grasp synthesis, commonly the shape of object, the orientation and the position are known [12]. Some other approach is focus on extracting the outer contour of an object and the feed back it to the planar grasp [13]. The work on contact-level grasp synthesis concentrates mainly on finding a fixed number of contact locations without considering the hand design [14, 15]. Considering hand kinematics and a priori knowledge of the feasible grasps has been acknowledged as a more flexible and natural approach toward automatic grasp planning [12].

Two kind of under-actuated robot finger, linkages and tendon-actuated mechanisms have been proposed in previous works. Linkage mechanisms are usually preferred for applications in which large grasping forces are expected. Hence, Tendon systems are generally limited to rather small grasping forces and they lead to friction and elasticity.

A valuable study on the statics of under-actuated grippers is presented in [16]. Two fingers with 4 DOFs are performed with statically analysis. It is also present the advantages of under-actuated fingers over a simple parallel gripper. Bartholet present an under-actuated hand with three fingers [17], which each finger consist of two phalanges and one actuator with 2 DOFs. Additionally, a special mechanism is introduced in order to allow the distal phalanges to be maintained orthogonal to the palm when precision grasps are performed. Crowder et al. present a mechanical

Fig. 1 Schematic drawing of
the proposed under-actuated
finger

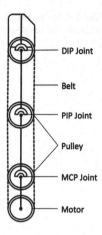

DIP Joint

Belt

PIP Joint

Pulley

MCP Joint

Motor

hand resembling the human hand in [18]. Each of the fingers is composed of three phalanges but has only 2 DOFs since the motion of the last phalanx is directly coupled to the motion of the second phalanx. In this chapter, we will present a new under-actuated robot finger with passive adaptive grasping. In design of these robot finger we take the advantage of under-actuation, i.e. the number of actuators are less than active joints, and combined with novel under patent mechanism for self-adaptive functionality.

3 Finger Mechanism

Based on resembling a healthy human hand, proposed finger consists of three pha-lanxes and three joints. An under-actuated method was adopted in order to drive all the joints of a finger by a single motor. A schematic drawing of the proposed under-actuated finger is shown in Fig. 1.

The structure of the finger can be simplified to a mechanism jointed at the Prox-imal Interphalangeal joint (PIP). Any relative motions of the middle phalanx to the proximal phalanx need to overcome the pretightening force of the pulley. The whole finger will keep straight while revolving around the metacarpophalangeal (MCP) joint until a proper force impedes the proximal phalanx. Next, The distal phalanx is rotating around the distal interphalangeal (DIP) joint while the middle phalanx is rotating around the PIP joint. A closing sketch of the proposed under-actuated finger is shown in Fig. 2.

4 Kinematic Model

Figure 3 shows the planar kinematic model of the robot finger in a 2D coordinate system. The parentheses are labeled according to the corresponding joints MCP, PIP and DIP of human finger.

Fig. 2 Closing sketch of
the proposed under-actuated
finger

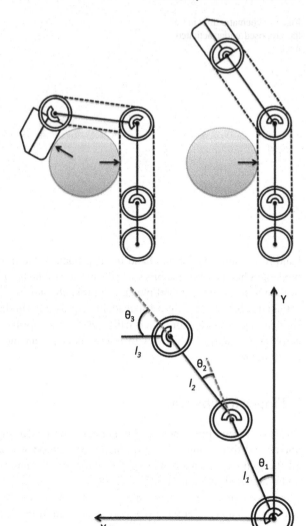

Fig. 3 Kinematic model of
robot finger

A solution of kinematic model of robot finger is presented in [19]. The rotation angle of the three joints is described by θ_i ($i = 1, 2, 3$). The length of links is l_i ($i = 1, 2, 3$), and initial position of the finger is assumed to extend in Z direction. The finger's tip position P is defined as

$$P = T_1(l_1 + T_2(l_2 + T_3 l_3))$$

where l_i is the link vectors that expressed by $l_i = (0, 0, l_i)^T$ and the rotational matrices T_i is expressed by

$$T_i = \begin{bmatrix} k_i & 0 & r_i \\ 0 & 1 & 0 \\ -r_i & 0 & k_i \end{bmatrix}$$

$$k_i = \cos\theta_i \quad r_i = \sin\theta_i \quad (i = 1, 2, 3)$$

Then, the obtained forward kinematics is

$$P = \begin{pmatrix} P_x \\ P_y \\ P_z \end{pmatrix} = \begin{pmatrix} l_1 r_1 + l_2 r_{12} + l_3 r_{123} \\ 0 \\ l_1 k_1 + l_2 k_{12} + l_3 k_{123} \end{pmatrix}$$

where the abbreviations are $k_{12} = \cos(\theta_1 + \theta_2)$, $K_{123} = \cos(\theta_1 + \theta_2 + \theta_3)$, $r_{12} = \sin(\theta_1 + \theta_2)$, $r_{123} = \sin(\theta_1 + \theta_2 + \theta_3)$.

A linear coupling factor is introduced as $\theta_2 = s\theta_3$ to express linear coupling of the angle. The forward kinematics equations can be described by introducing the coupling factor into Eq. 1.

$$P_x = l_1 \sin\theta_1 + l_2 \sin(\theta_1 + s\theta_3) + l_3 \sin(\theta_1 + (s + 1)\theta_3)$$
$$P_y = 0$$
$$P_z = l_1 \cos\theta_1 + l_2 \cos(\theta_1 + s\theta_3) + l_3 \cos(\theta_1 + (s + 1)\theta_3)$$

The planar geometry of a finger can be described by above equation with only two independent variables, θ_1 and θ_3.

Next, we calculate the angle θ_i from a given finger's tip position P. Firstly; the finger's tip position P can be rewrite as

$$T_1^{-1} P = l_1 + T_2(l_2 + T_3 l_3)$$

Expanding form can be describe as two nonlinear equations

$$P_x k_1 - P_z r_1 = l_2 r_2 + l_3 r_{23}$$
$$P_x r_1 - P_z k_1 = l_1 + l_2 k_2 + l_3 k_{23}$$
$$P_y = 0$$

One equation with two unknowns θ_2 and θ_3 described as

$$P_x^2 + P_z^2 = l_1^2 + l_2^2 + l_3^2 + 2l_1 l_2 k_2 + 2l_1 l_3 k_{23} + 2l_2 l_3 k_3$$

Then, the equation with one unknowns can be calculated by introducing the linear coupling $\theta_2 = s\theta_3$.

$$P_x^2 + P_z^2 = l_1^2 + l_2^2 + l_3^2$$
$$+ 2l_1 l_2 \cos s\theta_3 + 2l_1 l_3 \cos(s + 1)\theta_3 + 2l_2 l_3 \cos\theta_3$$

Generally, s is a real number and as pointed out in [20] it is mostly from 1.5 to 2 for human finger. The algebra solution doesn't exist when the irrational coupling ratio makes the inverse kinematics problem transcendental. Therefore, we limit it to rational coupling factor $s = M/N$, where M and N are natural numbers. Then the above equation can rewrite as

$$P_x^2 + P_z^2 = l_1^2 + l_2^2 + l_3^2$$
$$+ 2l_1l_2 \cos \frac{M}{N}\theta_3 + 2l_1l_3 \cos(\frac{M}{N} + 1)\theta_3 + 2l_2l_3 \cos \theta_3$$

Substitute $\theta' = \theta_3/N$ into above equation

$$P_x^2 + P_z^2 = l_1^2 + l_2^2 + l_3^2$$
$$+ 2l_1l_2 \cos M\theta_3' + 2l_1l_3 \cos(M + N)\theta_3' + 2l_2l_3 \cos N\theta_3'$$

A solution for solving this equation is mentioned at [19] and it can be express as

$$(P_z^2 - (l_2k_2 + l_3k_{23})^2) \tan^2 \theta_1 - 2P_x P_z \tan \theta_1$$
$$+ P_x^2 - (l_2k_2 + l_3k_{23})^2 = 0$$

This polynomial can solve by using formula for second order polynomial solution. A conclusion mentioned in [19] shows that it is desirable for a robot joint with an integer-coupling ratio of a human finger. So, using formulas for second or third order polynomial can solve the inverse kinematics.

5 Concepts and Design

Recently, most work of robot hand design is focus on ability to follow human hand structure and movement. All these works can categorize as three kinds of robot hands, mechanical grippers, specific purpose hands and universal hands. The first two type hands are designed and developed for working in special environment or for industrial automation. The universal robot hand is needed to be smooth grasping and flexibility to handle objects with different shapes and sizes.

To cover all these functions, the robot hand required designing with many Degrees of Freedom (DOF), which make complex design and complicated movement. So, it caused to increasing the weight, size, cost and consumption of energy.

In our design we focused on three concepts of environment, mechanism and control as described bellow.

Environmental concept: The robot finger with capacity of working under the daily life environment, grasping different shapes and material and grasping and lifting up the cans and bottles is required.

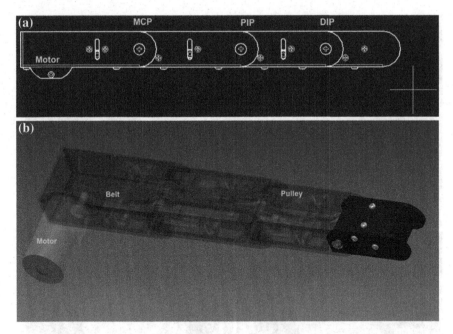

Fig. 4 2D and 3D sketch of the proposed under-actuated finger, **a** 2D sketch of proposed finger, **b** 3D sketch of proposed finger

Mechanical concept: All links and joints should drive by a single motor, which is placed in outside of finger. Also, the lengths between the joints should reference to the proportion of the human finger.

Control concept: The finger should design with minimal actuation and simple control without need for sensor feedback.

Based on above concepts and resembling a healthy human hand, proposed finger consists of three phalanxes and three joints. An under-actuated method was adopted in order to drive all the joints of a finger by a single motor. 2D and 3D sketches of the proposed under-actuated finger designed by Autodesk Inventor is shown in Fig. 4.

The structure of the finger can be simplified to a mechanism jointed at the Proximal InterPhalangeal joint (PIP). Any relative motions of the middle phalanx to the proximal phalanx need to overcome the pretightening force of the pulley.

The whole finger will keep straight while revolving around the MetaCarpoPhalangeal (MCP) joint until a proper force impedes the proximal phalanx.

Next, The distal phalanx is rotating around the Distal InterPhalangeal (DIP) joint while the middle phalanx is rotating around the PIP joint. A closing sketch of the proposed under-actuated finger is shown in Fig. 5.

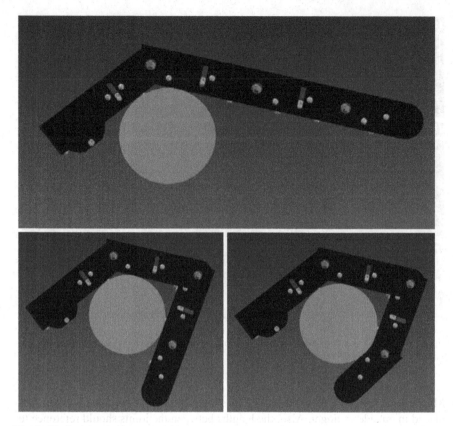

Fig. 5 Closing sketch of proposed finger

6 Under-Actuated Mechanism

The objective of the research is to design a flexible mechanical robot finger with capability of performing the grasping of a wide variety of objects. In our design, we take an advantage of under-actuated mechanism, which leads to shape adaptation of the finger. Under-actuated mechanism is a mechanical system which the number of control inputs is smaller than the Degrees of Freedom (DOF). This mechanism has many advantages such as low cost, low energy consumption and lightweight.

The concept of under-actuation leads to flexibility when applied to the mechanical finger. Flexible finger will envelope the objects to be grasped and self-adaptability to their shape with single actuator and without complex control.

The proposed finger takes the advantage of under-actuated mechanism and consists of three phalanxes and three joints. An under-actuated method is adopted in order to drive all joints of fingers by a single actuator. Figure 6 illustrate the main structure of proposed finger including the pulleys and the belts.

Fig. 6 3D sketch of proposed finger including pulleys and belts

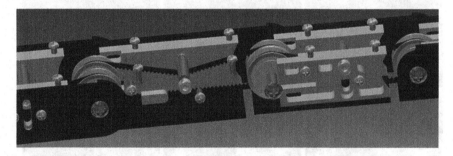

Fig. 7 MCP and PIP links

Figure 7 shows the MCP and PIP links in detail. MCP and PIP joints consist of two pulleys and one belt. The shaft in the center of each joint is fixed and controlled by patented attachments in both sides of joint.

Figure 8 illustrate the inner design of proposed finger including all links and joints from topside.

7 Flexibility of Proposed Design

An under-actuated mechanism is one that has fewer actuators than DOFs. By applying this mechanism into the robot fingers, the concept of under-actuation leads to flexibility. Flexible functionality fingers are envelope the objects to be grasp and automatically adapt to their shape with only one actuator without complex control. In statically under-actuated mechanisms elastic elements and mechanical limits must

Fig. 8 Inner design of proposed finger

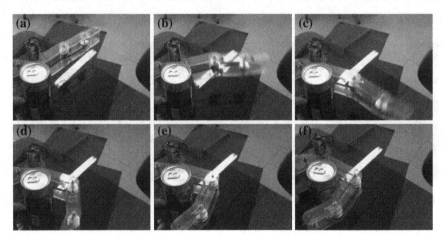

Fig. 9 Closing sequence

be introduced. Configuration of the robot finger is determined by the external constraints associated with the object while a finger is closing on an object. The force applied at the actuator is distributed among the phalanges when the object is fully grasped. A closing sequence of an under-actuated 3DOF robot finger is shown in Fig. 9.

This finger used one actuator and 3DOF and two elastic elements. In the example shown above, a mechanical limit is used to keep the phalanges aligned under the action of the pulley when no external forces are applied on the phalanges. Note that the sequence occurs with a continuous motion of the actuator. In (a), no external forces are present and the finger is in its initial position. The robot finger behaves as a single rigid body in rotation about a fixed pivot. In (b), the proximal phalanx and

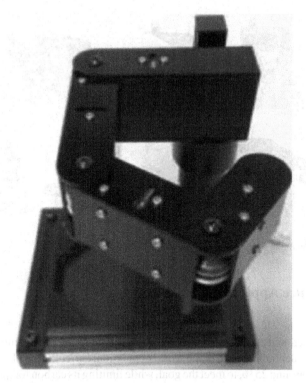

Fig. 10 The prototype of the proposed under-actuated robot finger

the object make contact. In (c), the second phalanx is moving with respect to the first phalanx.

The second phalanx is moving away from the mechanical limit. Then the finger is closing on the object since the proximal phalanx is constrained by the object. At this stage, the actuator has to produce the force required to extend the friction. Finally, in (f), phalanges are in contact with the object. Hence, the finger has completed the shape adaptation phase. The actuator force is distributed among the phalanges in contact with the object.

8 Experimental Results

The prototype of proposed under-actuated 3DOF robot finger is shown in Fig. 10. The size of robot finger is about $28 \times 240 \times 30$ (mm^3) and weight is nearly 420 (g) including pulley and belts. The MCP link size is 60 (mm), PIP link is 50 (mm) and DIP link is 45 (mm). The working voltage and current are 9 (V) and 1.8 (A).

The proposed under-actuated 3DOF robot finger can successfully operate various objects with different sizes and shapes based on the under-actuated mechanism of the finger (Fig. 11).

Fig. 11 Demonstration of flexibility of prototype robot finger

9 Agriculture Applications

Agriculture is the oldest and most important economic activity, which provide us food, fiber, feed and if necessary fuel for our basic needs and economic activities. Given the exponential increase of the world population, agriculture is also improving its efficiency, around 25 %, to meet the goal, while limiting its carbon footprint. Particularly in Japan, agriculture faces specific demographic challenges due to increasing number of elderly people and decreasing number of labor needed to continue agriculture projects. In this specific scenario robots become much more important in order to replace labor in the near future with a cost efficient and multi task solution.

Robotics and automation play a key role in agriculture, allowing price reduction and increasing the process efficiency. An example of that increasing role it is the market availability of different GPS guided harvesters and tractors as the one show in Fig. 12. Currently, there are several developments regarding specific robots developed for each process, i.e. pruning, thinning, and harvesting, as well as mowing, spraying, and weed removal. This kind of robots will help farmers to become more competitive in the market, but there is also expected to improve process efficiency.

In particular grasping tools are required to help improve farmer's efficiency and to improve harvesting time. Several complex models are being developed to solve this particular issue since each crop need a different way to handle due to its unique characteristics, i.e. Figs 13 and 14.

The proposed robot improves drastically the development time as well as its application; since it could self adapt its grasping characteristics when interacting with each crop, which makes this robot unique and useful to be used in several crops changing only its size, rather than create an specific application for each crop.

Fig. 12 Autonomous GPS based tractor

Fig. 13 MIT distributed robotics lab

10 Conclusions

A novel approach to mechanical design of under-actuated robot finger with passive adaptive grasping was presented in this chapter. Three concepts of environment, mechanism and control of passive adaptive grasping are also focused in our approach.

An under-actuated method is adopted in order to drive all joints of fingers by a single actuator. Also, the concept of under-actuation leads to flexibility when applied to the mechanical finger.

The proposed finger takes the advantage of under-actuated mechanism and consists of three phalanxes and three joints. Experimental results show proposed finger is enveloped the objects to be grasped and self-adaptability to their shape with single actuator and without complex control.

Future work will concentrate on validating the proposed finger by testing more objects with different shapes and materials.

Fig. 14 Institute of agricul-
tural machinery at Japan's
national agriculture and food
research organization

References

1. T. Laliberté, C.M. Gosselin, Simulation and design of underactuated mechanical hands. Mech. Mach. Theory **33**(1–2), 39–57 (1998)
2. R.O. Ambrose, H. Aldridge, R.S. Askew, R.R. Burridge, W. Bluethmann, M. Diftler, C. Lovchik, D. Magruder, F. Rehnmark, Robonaut: nasa's space humanoid. IEEE Int. Syst. Appl. **15**(4), 57–63 (2000)
3. J. Butterfass, M. Grebenstein, H. Liu, G. Hirzinger, DLR-Hand II: next generation of a dextrous robot hand, in *Proceedings of IEEE Internationall Conference on Robotics and Automation*, vol. 1 (Seoul, Korea, 2001), pp. 109–114
4. A. Bicchi, Hands for dexterous manipulation and robust grasping: a difficult road toward simplicity. IEEE Trans. Robot. Autom. **16**(6), 652–662 (2000)
5. C.Y. Brown, H.H. Asada, Inter-finger coordination and postural synergies in robot hands via mechanical implementation of principal components analysis, in *Proceedings of IEEE/RSJ International Conference on Intelligent Robots and System*, pp. 2877–2882, Oct. 2007
6. M. Ciocarlie, P. Allen, On-line interactive dexterous grasping. Springer Lect. Notes Comput. Sci. **5024**, 104–113 (2008)
7. T. Wimboeck, B. Jan, G. Hirzinger, Synergy level impedance control for a multifingered hand, in *Proceedings of IEEE International Conference on Robotics and Automation*, (Shanghai, China, 2011) pp. 973–979
8. D. Prattichizzo, M. Malvezzi, A. Bicchi, On motion and force controllability of grasping hands with postural synergies, In *Proceedings of Robotics Science and Systems*, Zaragoza, Spain, 2010
9. F. Cheraghpour, S. Moosavian, A. Nahvi, Multiple aspect grasp performance index for cooperative object manipulation tasks, in *Proceedings of IEEE/ASME International Conference on Advanced Intelligent Mechatronics*, pp. 386–391. July 2009
10. R. Murray, Z. Li, S. Sastry, *A Mathematical Introduction to Robotic Manipulation*, CRC Press, Boca Raton (1994)
11. C. Klein, B. Blaho, Dexterity measures for the design and control of kinematically redundant manipulators. The Int. J. Robot. Res. **6**(2), 72–83 (1987)
12. A.T. Miller, S. Knoop, H.I. Christensen, P.K. Allen, Automatic grasp planning using shape primitives, in *Proceedings of the IEEE International Conference Robotics and Automation*, vol. 2, pp. 1824–1829. Sept 2003
13. A. Morales, E. Chinellato, A.H. Fagg, A.P. del Pobil, Using experience for assessing grasp reliability. Int. J. Humanoid Rob. **1**(4), 671–691 (2004)
14. A. Bicchi, V. Kumar, Robotic grasping and contact: a review, in *Proceedings of IEEE International Conference on Robotics and Automation*, vol. 1, pp. 348–353. April 2000

15. D. Ding, Y.-H. Liu, S. Wang, Computing 3-d optimal form-closure grasps, in *Proceedings of IEEE Internationall Conference on Robotics and Automation*, vol. 4, pp. 3573–3578. 2000
16. H. Shimojima, K. Yamamoto, K. Kawawita, A study of gripppers with multiple degrees of mobility. JSME Int. J. **30**(261), 515–522 (1987)
17. S.J. Bartholet, Reconfigurable End Effector, U.S. Patent 5108140 (1992)
18. R.M. Crowder, D.R. Whatley, Robotic Gripping Device Having Linkage Actuated Finger Sections, U.S. Patent 4834 443 (1989)
19. I. Godler, K. Hashiguchi, T. Sonoda, Robotic finger with coupled joints: a prototype and its inverse kinematics, in *Proceedings of 11th IEEE International Workshop on Advanced Motion Control*, pp. 337–342. 2010
20. M.C. Hume, H. GGellman, H. McKellop, R.H. Brunfield, Functional range of motion of the joint of the hand. J. Hand Surg. **15A**(2), 240–243 (1990)

Discovering Unpredictably Related Words from Logs of Scholarly Repositories for Grouping Similar Queries

Takehiro Shiraishi, Toshihiro Aoyama, Kazutsuna Yamaji,
Takao Namiki and Daisuke Ikeda

Abstract As the number of institutional repositories is increasing, more and more people, including non-researchers, are accessing academic contents on them via search engines.   User models of non-researchers are not well understood yet, unlike researchers, although non-researchers may use quite different queries from researchers. For understanding their search behavior, it is a good way to categorize search queries of non-researchers into groups. This chapter is devoted to finding related query words at the first step from logs of scholarly repositories. In particular, we try to find words which are related from the viewpoint of non-researchers. In this sense, these words are unpredictably related. A simple method to do this using the access log is that we treat queries which lead to the same paper as related. However, it is challenging because one academic paper generally has a small amount of accesses while accesses to one paper bring many kinds of query words. Instead, we expand relationships between query words and papers, and use a graph-based algorithm in which query words and papers are vertices to find related words. As experiments, we use more than 400,000 accesses recorded at a major portal site of Japanese theses, and show that we can find related words with respect to specific disciplines if these words appear frequently. There words seems to be interested in non-researchers and hence

T. Shiraishi · D. Ikeda (✉)
Department of Informatics, Kyushu University, Fukuoka, Japan
e-mail: daisuke@inf.kyushu-u.ac.jp

T. Shiraishi
e-mail: 2IE12014P@s.kyushu-u.ac.jp

T. Aoyama
Department Electronic and Information Engineering, Suzuka National College of Technology, Mie, Japan

K. Yamaji
National Institute of Informatics, Tokyo, Japan

T. Namiki
Department of Mathematics, Hokkaido University, Hokkaido, Japan

R. Y. Lee (ed.), *Applied Computing and Information Technology*,
Studies in Computational Intelligence 553, DOI: 10.1007/978-3-319-05717-0_4,
© Springer International Publishing Switzerland 2014

we can't say they are not related in a usual manner. This result implicates that we can obtain related words if we enrich relationships between technical terminologies using background knowledge, such as dictionaries.

Keywords Institutional repositories · Access log mining · Randome walk · Hitting time · Query expansion

1 Introduction

Scholarly repositories are playing important roles for open access. As of Oct. 2013, we have more than 3,400 repositories in the world, according to Registry of Open Access Repositories.[1] Because major search engines began to index contents on repositories, many accesses to institutional repositories come from them [1, 7], and, moreover, some researches strongly indicate that non-researchers access contents on repositories [2, 6].

For example, people in Japan are more concerning about radioactive materials and their effects to people and environment, and so they are searching documents about radioactive materials and related topics. We can find such concerns in log data of repositories. In fact, repository managers at Otaru University of Commerce found that there exists some amount of accesses to a quite old paper published as a departmental bulletin chapter [5] after Fukushima nuclear disaster followed by the big earthquake in Japan.

Scholarly repositories seem to be basically designed for researchers but not for non-researchers because the original search model of repositories are based on meta data exchange using OAI-PMH and thus users are considered to use a harvester, which is a meta data search engine, such as OAISter.[2] Thus, it is important to understand behavior of non-researchers in order to provide relevant information for them at scholarly repositories.

For this direction, it is the first step to categorize search queries of non-researchers. To do so, in this chapter, we consider to find related query words, from the viewpoint of non-researchers, using log data. In this sense, these words are unpredictably related. In fact, we found that a user searches papers using the query with "sea urchin" and "radioactive material", which means the user, maybe a non-researcher, worries about the effect of radioactive material to sea urchin. In general, we can't say these two words are related, but, from the user's and some academic discipline's perspective, they are closely related, and thus we want to extract them as related words.

A simple method to do this using the access log is that we treat queries which lead to the same paper as related. This can be done by referer fields of accesses from search engines since referer fields of such accesses hold search queries given

---

[1] http://roar.eprints.org/

[2] http://www.oaister.org/

at search engines. It is, however, challenging to extract related words because one academic paper has a small amount of accesses while accesses to one paper bring many kinds of query words. We can't use methods to find similar words that could be used for our purpose, such as a method in [4], because we need related words. We can't also use methods which require many training examples, such as topic models of the natural language processing, because of the data sparseness.

Instead, we use a graph-based algorithm, where we consider query words and papers constitute a bipartite graph. More precisely, we use the algorithm in [3], which was originally proposed for query expansion. We use this because we can add edges using background knowledge even for a sparse graph.

The effectiveness of the method is evaluated using real log data of more than 400,000 accesses. We found some interesting related words with respect to specific disciplines if these words appear frequently. For example, we obtained that "the elderly" and "bathing period" are related through the constructed graph. These are not similar but related because there exists a close relationship between health-care for the elderly and bathing period.

2 Bipartite Graph and Hitting Time

In this section, we introduce some mandatory notions related to a bipartite graph and a hitting time used in our method, according to [3].

A bipartite graph is a graph with two types of vertices, and edges between them showing the relationship among two different vertex groups. In our case, two types correspond to queries and theses, respectively. A hitting time is the first time at which a given random process reaches a node in a given subset of nodes, and it is used to estimate the expectation of reaching time to any vertex in the subset from one vertex.

Formally, these are defined as follows. A bipartite graph is a graph $G = (V, E)$ in which there exists a partition $V = V_1 \cup V_2$ such that every edge in E connects a vertex in V_1 and one in V_2, and there is no edge between two vetices in the same set, that is, $a \in V_1$ is not connected to $b \in V_1$. Let $w : V_1 \times V_2 \to R^+$ denote a weight function such that, given $i \in V_1$ and $j \in V_2$, $w(i, j)$ is positive if there is an edge between i and j, and $w(i, j) = 0$ otherwise.

A random walk on a bipartite graph can be defined as follows. Assume the current position is at a vertex in V_1. Then an edge connected to this vertex is chosen with the probability to the weight of the edge. By following this edge, the random walk arrives at a vertex in V_2. Then, as well as, an edge connected to V_2 is chosen to follow and the random walk goes back to V_1. Therefore, given $i \in V_1$ and $j \in V_2$, the transition probability is defined as

$$p_{ij} = \frac{w(i, j)}{d_i}, \tag{1}$$

where $d_i = \sum_{j \in V_2} w(i, j)$. If we focus only in the vertices in one side, such as V_1, then a new random walk in a query to another query is

$$p_{ij} = \sum_{k \in V_2} \frac{w(i,k)}{d_i} \frac{w(k,j)}{d_k}. \qquad (2)$$

It is obvious that stationary probability π_i is proportional to d_i. The above portion is the description of the random walk in a bipartite graph $G = (V, E)$. And in what follows, we will discuss hitting time on a graph $G = (V, E)$.

Let A be a subset of V and X_t denote the position of the random walk at discrete time t. We define the hitting time T^A is the first time that the random walk is at a vertex in A, that is, $T^A = min\{t : X_t \in A, t \geq 0\}$. It is obvious that T^A is a random variable. Given $i \notin A$, we obtain from the definition of the hitting time,

$$P[T^A = m|X_0 = i] = \sum_{j \in V} P[X_1 = j|X_0 = i]P[T^A = m - 1|X_0 = j]$$

$$= \sum_{j \in V} p_{ij} P[T_A = m - 1|X_0 = j]. \qquad (3)$$

Let h_i^A denote the expectation of T^A under the condition $X_0 = i$, and $h_i^A = E[T_A|X_0 = i]$. Then, we have

$$h_i^A = \sum_{m=1}^{\infty} m P[T^A = m|X_0 = i]$$

$$= \sum_{m=1}^{\infty} m \sum_{j \in V} p_{ij} P[T^A = m - 1|X_0 = j]$$

$$= \sum_{j \in V} \sum_{m=1}^{\infty} (m - 1) p_{ij} P[T^A = m - 1|X_0 = j]$$

$$+ \sum_{j \in V} \sum_{m=1}^{\infty} p_{ij} P[T^A = m - 1|X_0 = j]. \qquad (4)$$

For the first term, we have

$$\sum_{j \in V} \sum_{m=1}^{\infty} (m - 1) p_{ij} P[T^A = m - 1|X_0 = j]$$

$$= \sum_{j \in V} p_{ij} \sum_{m=1}^{\infty} m P[T^A = m|X_0 = j]$$

$$= \sum_{j \in V} p_{ij} h_j^A, \qquad (5)$$

and, for the second term, we have

Fig. 1 The concept of mean hitting time

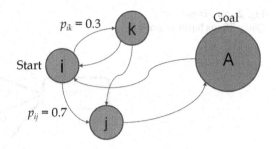

| A : a subset of V |
| X_t : a position at time t |
| i : start query |
| j, k : query |
| p : transition probability |

$$\sum_{m=1}^{\infty} P[T^A = m - 1 | X_0 = j] = 1, \tag{6}$$

and so

$$\sum_{j \in V} \sum_{m=1}^{\infty} p_{ij} P[T^A = m - 1 | X_0 = j] = \sum_{j \in V} p_{ij} = 1. \tag{7}$$

Consequently,

$$h_i^A = \sum_{j \in V} p_{ij} h_j^A + 1. \tag{8}$$

Note that $h_i^A = 0$ for $i \in A$.

Combining all above pieces together, we obtain the linear system for computing the mean hitting time which is an average of hitting time as follows:

$$\begin{cases} h_i^A = 0 & \text{for } i \in A \\ h_i^A = \sum_{j \in A} p_{ij} h_j^A + 1 & \text{for } i \notin A \end{cases}$$

For example, in Fig. 1, we can compute the mean hitting time $h_i^A = 0.7 h_j^A + 0.3 h_k^A + 1$. And this value is going to be more precise with the iterative computing. As we mentioned the hitting time is the first time that the random process reaches a node in a given subset, more shorter hitting time indicates that two queries have more stronger relationship. And the mean hitting time is the average of the hitting time, therefore if mean hitting time is short, which means that the semantic distance of two queries are close. By using this feature we can extract in descending order of related queries for the target query.

Fig. 2 An example of a
Query-Thesis bipartite graph

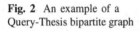

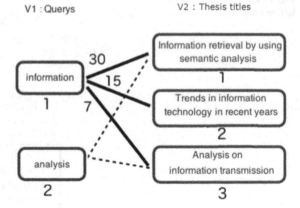

3 Discovering Related Words Using the Hitting Time

In this section, we explain how to implement the linear system we just described for
log data of repositories.

We assume that we extract pairs of the form, $\langle Q_i, T_i \rangle$, from the log data, where
Q_i represents a query and T_i a thesis. That is, a user gave Q_i as a query at a major
search engine and clicked T_i.

Using these pairs, we construct a bipartite graph $G = (V, E)$, where $V = V_1 \cup V_2$,
V_1 corresponds to all queries, and V_2 all theses. Each edge $e = (i, j) \in E$ corresponds
to a pair $\langle Q_i, T_j \rangle$ with a positive frequency which shows how many times this pair
appears in the log data. We weight each edge with $w(i, j) = C(Q_i, T_j)$, which is
the number of accesses to T_j with Q_i.

A simple example of an undirected Query-Thesis bipartite graph is shown in Fig. 2.

We see that every query is connected with a number of theses on which users
clicked after submitting corresponding queries. The weights on the edges denote
how many times the users used this query to access this thesis. Note that there is no
edge connecting two queries, or two theses.

For example some people did the search activity with the query "information".
Then the search activity reached to the thesis "Information retrieval by using semantic
analysis". And the such behavior have been observed 30 times in the logs of scholarly
repositories. Exactly, the set of the query "information" and the thesis "Information
retrieval by using semantic analysis" appeared 30 times in the access log. The meaning of weight on the top edge in Fig. 2 is so. As well as we can comprehend the
meaning of weight on other edges in the same way. That is the set of the query
"information" and the thesis "Trends in information technology in recent years"
appeared 15 times, and the set of the query "information" and the thesis "Analysis
on information transmissioin" appeared 7 times in the access log.

We can compute the query-thesis transition probability from this bipartite graph.
For example, p_{11} in Fig. 2, which is the transition probability of "information" to

Fig. 3 An example of a
Query-Query bipartite graph

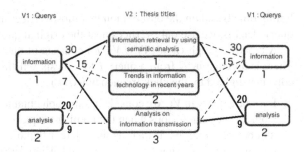

"Information retrieval by using semantic analysis" is computed as $p_{11} = \frac{30}{52}$. It is possible to also compute the following transition probabilities p_{12} and p_{13}, in the same manner, that is, $p_{12} = \frac{15}{52}$ and $p_{13} = \frac{7}{52}$. That is, we define a transition probability from the query to the thesis depending on the number of queries reaches the paper in the access logs.

If we focus on only queries (vertices in V_1), it is possible to determine the transition probability from query to another query. Figure 3 is an example of a graph that reaches query through thesis from another query. As we can see it's just an extension of the Query-Thesis bipartite graph. This graph is considering not only the transition from the query to the thesis but also the transition from the thesis to the query again. Please note that the direction of edges are reversed, but the weight of each edge is the same as in the previous Query-Thesis graph.

Now we can compute the query-query transition probability from this bipartite graph. For example, p_{12} in Fig. 3, which is the transition probability of "information" to "analysis" is computed as $p_{12} = \frac{30}{52}\frac{20}{50} + \frac{7}{52}\frac{9}{16}$. That is the transition probability of "information" to "analysis" equals the value of the edge from the query "information" to the thesis "Information retrieval by using semantic analysis" times, the value of the edge from the thesis "Information retrieval by using semantic analysis" to the query "analysis" plus, the value of the edge from the query "information" to the thesis "Analysis on information transmissioin" times the value of the edge from the thesis "Analysis on information transmission" to the query "analysis". In this way, we obtain the transition probability to all queries that are connected through all theses.

Using the Query-Query bipartite graph, we can suggest more widely words. Because the Query-Query bipartite graph is able to consider the relationship between the two queries even the two queries are not connected directly with one thesis. In contrast the Query-Thesis bipartite graph focuses only queries are connected directly with one thesis.

Let Q_T be the target query. With the setting of $A = \{Q_T\}$, we can compute the hitting time $h^A(i)$ for all other queries Q_i based on the bipartite graph, use this measure to rank Q_is, and select the top-k queries as same group to Q_T.

Now we can compute the hitting time using the constructed graph. However, the size of the graph is quite large in general and thus it requires a huge amount of computational resources. Therefore, we use some heuristics when computing the hitting time based on the observation that most vertices are irrelevant to the original query. That is, we use only vertices near the original query.

A bipartite graph $G = (V_1 \cup V_2, E)$ consists of query set V_1 and thesis set V_2. There is an edge in E from a query i to a thesis k if this thesis is clicked, and the edge is weighted by the click frequency $w(i, k)$.

1. Let s be a query in V_1. We consider a subgraph obtained by depth-first search in G from s. The search stops when the number of queries is larger than a predefined number of n queries. If there is more larger depth of depth-first search, queries recommended as candidates are increased. Therefore, computing time and depth is a trade-off, and we will be required to set appropriate depth on a basis of experience. This time, we have with depth $= 2$. It means we follow twice the reachability to the query from another query.
2. We form a random walk on the subgraph by defining transition probabilities between two queries i and j in V_1 as

$$p_{ij} = \sum_{k \in V_2} \frac{w(i, k)}{d_i} \frac{w(k, j)}{d_k}. \tag{9}$$

3. For all queries except the given one (query s), we compute the following iteration

$$h_i(t + 1) = \sum_{j \neq s} p_{ij} h_j(t) + 1 \tag{10}$$

for a predefined number of m iterations started with $h_i(0) = 0$. Usually, by performing the iteration of 30 times, hitting time is converge. This time, we have with $m = 30$.
4. Let h_i^* be the final value of $h_i(t)$. And output the queries which have the top k smallest h_i^* as suggestions. This time, we have with $k = 5$.

4 Experiments

In this section, we use our experimental results to show the effectiveness of the proposed algorithm.

4.1 Data Set

We collect a large scale query log dataset of CiNii,[3] which is a major portal site of papers in Japan, among about 8 days from December 1, 2008 to December 8, 2008.

---

[3] http://ci.nii.ac.jp/

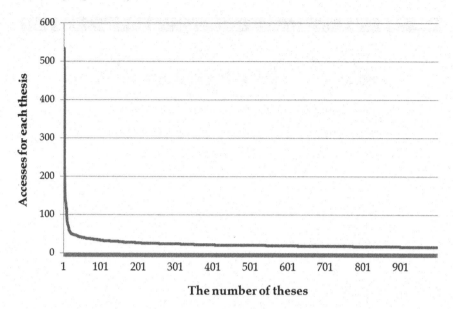

Fig. 4 A distribution of the number of accesses to each thesis

IP adress   Date   Thesis id
118.155.241.135 - - [01/Dec/2008:08:59:59 +0900] "GET /naid/110003475329/ HTTP/1.1" 200 27050 885520
                                                                      Query
"http://www.google.co.jp/search?hl=ja&rls=ADBR%2CADBR%3A2006-50%2CADBR%3Aja&q=Moving+Obstacle&btnG=%E6%A4%9C%E7%B4%A2&lr=lang_ja"
"Mozilla/4.0 (compatible; MSIE 6.0; Windows NT 5.1; SV1; InfoPath.1; .NET CLR 2.0.50727; .NET CLR 3.0.04506.30)"

Fig. 5 A sample access in CiNii log

This data contains 240,430 unique queries, and 286,633 unique theses. We show the distribution of the number of accesses to each thesis in Fig. 4. As we can see, it's really sparse data. By using the Query-Query bipartite graph, we try to recommend a wide range of queries even the number of access of logs is low.

CiNii log data has information of IP address, access time, referer (some of those have information reference thesis id and queries). We show a sample access recorded in CiNii log (see Fig. 5). For example in Fig. 5, the IP address is 118.155.241.135, the access data is December 1st 2008, the query is "moving obstacle", the thesis id is 110003475329. It means some people did the search activity with query "moving obstacle" at December 1st 2008 by 118.155.241.135. Then the search activity reached to the thesis that have the thesis id 110003475329. And we extracted lines of log data that have both information reference thesis id and queries like Fig. 5.

Then we obtain the queries by using URL decoding if it is necessary, and we obtain thesis titles by html tag throwing thesis id to the pages of the thesis information on CiNii home page. An example of thesis information page on CiNii is as shown in Fig. 6. As we mentioned CiNii is the academic portal site, it holds the author information and the thesis title etc as we can see from Fig. 6. In fact html is working in its back ground which links the thesis title and thesis id. Based on these information we create the set of the query and thesis.

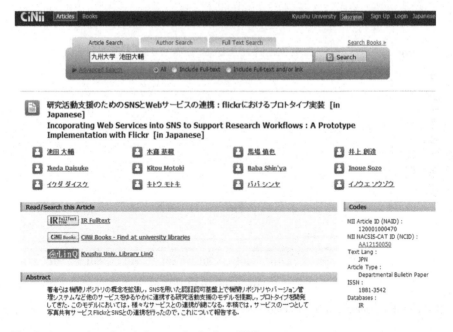

Fig. 6 A screenshot of a splash page of a thesis at CiNii

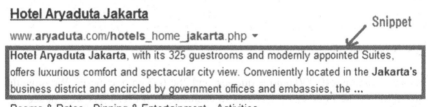

Fig. 7 A sample of the snippet

4.2 Naive Methods

Before we use the hitting time, we considered using a web corpus known as open data and already famous. The first method is using a categorization of Wikipedia, and the second method is obtaining a snippet from the search engine (In this case, we used Yahoo). We conducted these two experiments as naive methods to discovering the related words.

As the former method, we get the category the query belongs by throwing a query to Wikipedia. Then, if some queries belong to the same category, we put them into the same category.

For the latter, we get search results by throwing a query to Yahoo, and we get a snippet (summary statement appears at the bottom of the link for each search result) of the top 20. We show the sample of the snippet in Fig. 7.

Table 1 Grouping queries by using existing web corpus methods	Categories from Wikipedia	Words from Yahoo Snippets
	Query = The elderly	
	The elderly	Human
	Age	Old
		Information
		Age
		Employment
	Query = gene	
	Genetics	Research
	Molecular biology	Experiment
	Gene	Facility
		Congress
		Center
	Query = sound	
	Genre of music	Music
	List of music	Information
	Music	Voice
	Musical theory	Use
		Hearing

As we can see if the original query is "Hotel Aryaduta Jakarta" which is one of the most high-class hotels in Jakarta, we can get some words "luxurious" and "comfort" etc. Like this way we try to extract features or keywords to the original query. Then, we obtain the top 5 words that are most commonly used in the snippets, and we regard them are the same group of original query. We show results of these two methods in Table 1.

For the results from both methods, there is too wide range to grouping, and too general. Result words being aware of all web users are mostly, so we can not find leads scholarly relationship what we wanted. As we mentioned non-researchers are not familiar with the academic research. Therefore they don't know the close words have significant relationship in the research area. So we want to suggest queries with a strong connection even it's difficult to predict from the original query. However the results of naive methods don't meet the criteria to suggest query, because these suggested queries can be easily associated from the original query.

4.3 Using the Hitting Time

In order to solve the above problems, we adapted the hitting time to CiNii log data. As pre-processing, we have done the following: First, as mentioned in subsection 'Data set', we extracted session logs that have query and thesis information from the entire log. Then we created a Query-Thesis bipartite graph and Query-Query bipartite

Fig. 8 A flow of discovering
the related words

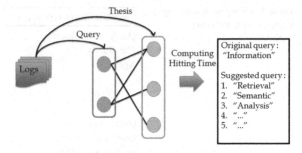

Table 2 Query suggestion for same group using hitting time on CiNii log data

Suggestion by hitting time	Suggestion by Google	Suggestion by Yahoo
Query = The elderly		
Bathing period	Recreation	Recreation
Mall order	Dehydration	Train training
Test	Ruiz	Gymnastics
Trachea	Feature	Illustration
Efficacy	Heat illness	Reason for living
Query = Gene		
gpx	Inspection	dna
Pen	Diet	dna difference
Receptor	dna	Research
White	News	Illness
Plural	Chromosome	Use
Query = Sound		
Based	Material	Not come out
Level	Frequency	Edit
Modulus	Spee	Frequency
Recognition	Material free	Materials
Class	Fastness	Mean

graph from that data. We computed hitting time on each queries from bipartite graph subjected to depth-first search. Finally, we get the top 5 queries for grouping with the original query. We show the flow of discovering the related words by using the bipartite graph and the hitting time in Fig. 8.

We show the comparison of suggestions generated from our algorithm and those from Google and Yahoo. We show that result on Table 2. Clearly, we can see that the suggestions generated with hitting time focus on relationship in the research field than suggestions currently provided by the two search engines.

As a result of this method, e.g., query "bathing period" is obtained for query "the elderly". This result indicates that there is a close relationship between bathing period and health status of the elderly people. And this also indicates that the users searched in order to find out this information from reliable research papers. As described above, that result as an index to know measure the interest of the public, and it is also a measure to know the movement in the research industry today. In contrast it has not been proposed only in general terms to the original query by using Google suggestion and Yahoo suggestion.

We will look at the results for the "gene". The first suggested query "gpx" is a kind of protein, and it has attracted attention in recent years as prevent oxidative damage. Oxidative damage is human cells being oxidized, and it seems as a kind of cause of cancer and human aging. Therefore, this result shows the trend of recent research such as Anti-Aging and cancer prevention.

For the results of the "sound", there are "recognition" and "level". These seems to relate to speech recognition. Other search engines focus on how to use the "sound", e.g., "material free", "edit". But the result of using hitting time focuses on the study of the sound itself, in a deeper level as further.

In this way, we can extract of query groups with the connection in research field, even the queries completely seem to be different at first glance. Basically non-researchers are not able to find related words in the academic area. Because these queries are unpredictable from the original query. But using this method they are able to find some relevant words in the academic area.

5 Conclusion

In this chapter, we proposed a method to find related words with respect to academic disciplines. Implementing the hitting time to CiNii log data, we succeeded in extracting a query group having a connection that is reliable in academic. At first glance, although they appear to be unrelated, closely related to the field of study.

As we mentioned earlier, accesses by non-researchers are increasing, so we guess these log data in CiNii also contain a lot of access by non-researchers. Although these might also include accesses by researchers somewhat, as important point, finally we obtained like a dictionary in the academic area and it is going to be helpful information when non-researchers are searching some papers.

As the current problem, if the edge of Query-Thesis is only one for each vertex, hitting time algorithm can not make a recommendation for another queries. Actually, CiNii is a paper repository and each paper is too technical. So it is very sparse corpus as a whole. Therefore, our future work is improved coverage for sparse query log.

In order to do so, we are considering one solution. That is increasing edges in a bipartite graph by using other information. For example if the two queries have same meaning in the academic area and only one query have the edge to one thesis, we are going to increase one edge from another query to the thesis. Then, the hitting time algorithm can make suggestion if the edge of original Query-Thesis graph is only one for each vertex.

We guess that can be done a grouping of appropriate query for the sparse log data, and be something to fit in the place that will be part of the tail in a normal web search not only in CiNii log data.

Further, to increase the accuracy of the grouping we have to discuss the depth of depth-first search. In this time, we defined the depth is 2 from the relationship of computing time. However, it is necessary to we inspect how enough precision is provided if we make which depth. So we will make it validate in the future.

Acknowledgments This work was supported by JSPS KAKENHI Grant-in-Aid for Scientific Research (B), Number 23300087.

References

1. D. Ikeda, S. Inoue, Access flows to a repository from other services, in *Proceedings of the 4th International Conference on Open Repositories* (2009),http://hdl.handle.net/1853/28422
2. D. Ikeda, P. Wang, Revealing presence of amateurs at an institutional repository by analyzing queries at search engine, in *Proceedings of the 7th International Conference of Open Repositories* (2012)
3. Q. Mei, D. Zhou, K. Church, Query suggestion using hitting time, in *Proceedings of the 17th ACM Conference on Information and Knowledge Management*, pp. 469–478 (2008)
4. M. Sahami, T.D. Heilman, A web-based kernel function for measuring the similarity of short text snippets, in *Proceedings of the 15th International Conference on World Wide Web*, pp. 377–386 (2006)
5. K. Saito, On the radiological contamination of environment by utilization of the atomic power. The Review of Liberal Arts **26**, 103–127 (1963). (in Japanese)
6. S. Sato, Y. Nagai, T. Koga, K. Misumi, H. Itsumura, How does article deposition in institutional repositories affect both citations and e-journal usage. J. Jpn. Soc. Inf. Knowl. **21**(3), 383–402 (2011). (in Japanese)
7. S. Sato, M. Yoshida, Usage log analysis of articles in six japanese institutional repositories: which region do users access articles from? in *The 2010 CiSAP colloquium on Digital Library Research, Doctoral Student Forum* (2010)

A Trichotomic Approach to Concept Capture and Representation: With its Application to Library Data Mining

Toshiro Minami, Sachio Hirokawa, Kensuke Baba and Eriko Amano

Abstract The aim of this chapter is twofold. Firstly, we propose a method of specifying the concept that is too hard to describe in an exact way by a word or a phrase, by setting up the "relative distances" from three key concepts; which we call a trichotomic approach to concept capture and representation, or description, in an approximate means. It is important and interesting that we can choose not only the key words but also other three "keys" such as patrons, books, concepts, objects or others. Then we arrange the objects of study according to the relative distances from these three keys, and investigate how these objects are distributed. Secondly, we demonstrate the usefulness of trichotomic approach through a couple of case studies applied to library's loan record analysis. In these case studies, we discuss and compare the methods of choosing three keys, then we show how the trichotomic representation method is applied to the real data analysis. From these case studies, we are convinced of its high potential and importance as a visualization tool of the results of data analysis in general.

Keywords Library marketing · Library data analysis · Loan/Circulation records · Concept representation · Trichotomic/Triangular representation

T. Minami (✉)
Kyushu Institute of Information Sciences, 6-3-1 Saifu, 818-0117 Fukuoka, Japan
e-mail: minami@kiis.ac.jp

S. Hirokawa
Research Institute for Information Technology, Kyushu University, Fukuoka, Japan
e-mail: hirokawa@cc.kyushu-u.ac.jp

K. Baba
Research and Development Division, Kyushu University Library, Fukuoka, Japan
e-mail: baba.kensuke.060@m.kyushu-u.ac.jp

E. Amano
Office for eResource Services, Kyushu University Library, Fukuoka, Japan
e-mail: amano.eriko.760@m.kyushu-u.ac.jp

R. Y. Lee (ed.), *Applied Computing and Information Technology*,
Studies in Computational Intelligence 553, DOI: 10.1007/978-3-319-05717-0_5,
© Springer International Publishing Switzerland 2014

1 Introduction

In this chapter, we study how to capture and represent various concepts based on the relative distances from 3 keys, which has several background motives, as we mentioned in [6]. The first motive comes from a study of investigating a method of expressing concept which has no appropriate words or phrases. In the study on ZK (ZeichenblocK, or ZakKi-chou in Japanese; meaning a note-pad) system [2], we proposed a method of specifying such a concept by giving a position based on the relative distances from known words.

Next motive comes from the observation that anomatopoeia has a powerful functionality in expressing nuances of a concept, i.e. word(s) and expression(s). According to a TV program [7], onomatopoeia is even used in a development of commercial product as well as it is used in commercial promotions to attract the consumers. Both ZK and use of onomatopoeia are common in the sense that they are for describing the concepts based on the keys (words, phrases, etc.). Because they have different roles in expressing non-standard concepts, they are somewhat complementary rather than competitive, in the sense we can describe a concept in our mind more exactly by mixing them up.

Library marketing is another motive of this study. Marketing becomes more and more important for libraries because of the advancement of information age and their patrons' needs has been vigorously changing. One of the most important topics for them is obviously to recognize what their patrons are. We have been conducting a series of studies of investigating the patrons' behavior by analyzing the data obtainable by libraries, mainly using the loan records intending to develop tools useful for library marketing (e.g. [3–6]).

We have a problem in our study in library data mining for marketing. It is on finding a good method for visualization of the results obtained from the data analysis of library data. As a possible solution to this problem, we present a trichotomy-based representation method. In this chapter we investigate the potentiality of using trichotomic approach as a new type of analysis methodology of library data; specifically of loan records. In the case studies we use a collection of loan records and we can choose not only the key words but also the patrons or faculties as the "keys" and set the location of them according to the "relative distances" from the three keys. In such a way, we can compare different types of objects such as patron and books in a uniform framework of trichotomic representation. This framework of distance can provide with another types of methods for data analysis and finding. Even though we show the methods using a collection of library's loan records, the methods themselves are more general and thus be able to apply to wider cases.

The rest of this chapter is organized as follows: In Sect. 2, we introduce some results of studies for library data analysis we have conducted so far. Also we discuss our approach to data analysis with emphasizing the importance of visualization. Then in Sect. 3, we deal with trichotomic approach to data analysis. Our approach in this chapter on this issue is to use of triangular representation of objects as well as concepts based on the relative distances from three key concepts/objects. We show a couple of case studies and demonstrate the usefulness of trichotomic approach in data analysis. Finally in Sect. 4, we summarize the discussions we had in this chapter and anticipate the research topics for the future studies.

2 Library Data Analysis

The data we use in this chapter are the book loan/circulation records obtained from the Central Library of Kyushu University, Japan, from April 2007 to March 2008, which were also used in [3–6]. The original data contain 67,304 records. A record consists of the book ID, book's Nippon Decimal Classification (NDC) number, call number, borrower's patron ID (which are renumbered by considering privacy issue), affiliation, type (B1–B6 for undergraduate, M for masters, and D for Ph.D students, P for academic staff or professors, and O for other patrons), and the timestamps for borrowed and returned dates and times, etc.

The borrowers affiliate either one of the 12 faculties or other organizations. The faculty names stand for; AG for Agriculture, DD for Dental, DS for Design, EC for Economy, ED for Education, LA for Law, LT for Letter, MD for Medicine, NC for the special faculty called twenty first century program, which is for the students who wish to study in a wide variety of subject fields, O for those in other organizations, PS for Pharmaceutical, SC for Sciences, and TE for Engineering.

We define the concept of patron's interest area profile in the same way as in the papers [4–6]. For the areas of topics, we use the NDC category numbers assigned to the books as a part of their bibliographic data. NDC is a Japan-localization of the decimal classification system Dewey Decimal Classification (DDC). The top level categories of NDC consist of the following 10 topic fields; 000 for General Works, 100 for Philosophy and Religion, 200 for History and Geography, 300 for Social Sciences, 400 for Natural Sciences, 500 for Technology (Engineering), 600 for Industry and Commerce, 700 for Arts, 800 for Language, and 900 for Literature.

We define the profile of a patron as a vector with dimension 10, each element corresponds to a top category of NDC. An element of the vector is the ratio of the number of books of the specified category which are borrowed by the patron. We extend this definition to a group of patrons by changing the condition from "borrowed by the patron" to "borrowed by one of the patrons of the group". We define the range of interest area of by the information entropy of the profile vector: Let p_i be the ratio of the books of NDC category i of the patron's profile for $i = 000$–900. Then the range of the profile is defined by $\sum_i -p_i \log_{10} p_i$. We use 10, the number of NDC categories, as the base of the logarithm so that the maximum value becomes 1. We use the cosine similarity of two profiles as the similarity measure in this chapter.

Figure 1 shows the interest area profiles of the top 11 patrons, called from P.A to P.K, according to the order of numbers of borrowed books. The figure shows that the ratios of books vary widely from patron to patron. For example, P.A borrowed from quite a wide area of books with NDC numbers from 000 to 900. On the other hand, P.C borrowed mostly from the classification number 400 (Natural Sciences).

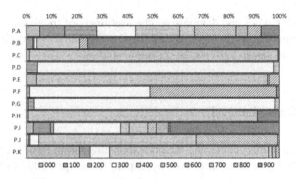

Fig. 1 Profiles of top 11 students

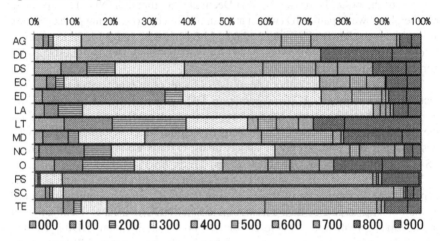

Fig. 2 Profiles of faculties

Figure 2 shows the profiles of faculties. We can see that Sciences (SC) and Pharmaceutical (PS) are very different from other faculties in their profile patterns. They have the two lowest range values, which mean that patrons, i.e. students, affiliated in them borrow the books mostly from natural sciences (NDC 400). Further the number of the borrowed books by SC students is also quite high, which probably comes from the fact that the SC building is located very close to the central library.

The faculties of DD (Dental) and LA (Law) also have small range values (about 0.45), which means that their student patrons also borrowed books mainly from their expertise subject areas than other faculties. It is interesting to see that Design (DS) and Medical (MD) have relatively high range values (>0.8). The MD students seem to visit the library from different campus in order not to find the books relating to their expertized field but to search for books in a wide variety of subjects. DS relates to both engineering and arts and should have a wide interest range as a whole. However it is still surprising that its range is larger than any other faculties including Other, or unclassified (O) and that Letter (LT) also has high range. The members of LT borrowed not just the books of literature (NDC 900), but also those in other areas as many as of literature.

3 Trichotomic Approach to Capture and Representation of Concept

In this section, we discuss how to deal with capturing concepts with trichotomic approach. Firstly in Sect. 3.1, we investigate how to represent the target concept and object in a trichotomic method. We present two types of representation methods; (1) by relative values for three pairs, and (2) by using triangular representation. Then in Sect. 3.2, we discuss the methods of choosing 3 keys (key words, key concepts, key objects, etc.). We show case studies in the remaining sections. These case studies aim not only to show the analysis results of library's loan records, but also to show the different methods of choosing 3 keys.

3.1 Triangular Representation of Trichotomic Concepts/Objects

In this section, we start with studying representation methods together with interfaces for specifying and showing the positioning of concepts/objects with relative distances from three keys (i.e. key words, key objects, or three whatever). Then we investigate theoretical conditions for locating a point that represents the relative distances from 3 keys.

Figure 3 shows a sample image of the interface in two types. Let A, B, and C be the keys. The left part (a) shows an interface design for specifying pairwise ratios. As we have three keys, we have three pairs; namely A and B, B and C, and C and A. Even with three pairs, we have only two degrees of freedom. Thus we set up three radio buttons at the left. The selected button indicates that this pair is dependent to other two. In Fig. 3a, the button for the pair C and A is marked as dependent, thus we can specify the ratios for the pairs A and B, and B and C.

For each pair, a slide bar is given and we can specify how the concept/object in mind is located in comparison with two keys. In Fig. 3a, it is given by the location of rather closer to B than A for the pair of A and B, with the ratio of $p : p'$ where $p' = 1 - p$ and $0 \le p \le 1$, so $0 \le p' \le 1$, and near the middle for the pair B and C, with $q : q'$ ($q' = 1 - q$), so that $0 \le q, q' \le 1$. As the result it is located closer to C than A, with $r : r'$ for the pair C and A.

The right part (b) of Fig. 3, the concept/object is shown as a location in an triangle with three vertexes A, B, and C. The location of the point G for the target concept is determined as the intersection point of line segments CC' and AA', where C' and

Fig. 3 Two types of representations for trichotomic distance ratios in: **a** pairwise interface and **b** triangular interface

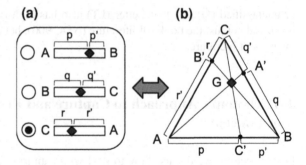

A' are the points that divide the side AB so that $AC' : C'B = p : p'$ and the side BC so that $BA' : A'C = q : q'$, respectively. The value of r for the ratio for C and A is calculated as follows: By Ceva's theorem, we have $pqr = p'q'r'$. By using the definition $r' = 1 - r$ we have $pqr = p'q'(1 - r) = p'q' - p'q'r$, so we have $(pq + p'q')r = p'q'$. As the result, we have the following two formulas.

$$r = \frac{p'q'}{pq + p'q'} \qquad r' = \frac{pq}{pq + p'q'}$$

By using these formulas, we can calculate r and r' from p and q and the position marker for the pair CA like in Fig. 3a shows these values of r and r'. For example, let $p = 0.7$ and $q = 0.6$, thus $p' = 0.3$ and $q' = 0.4$. Then we have the following value; which case is shown in Fig. 3.

$$r = \frac{0.3 \times 0.4}{0.6 \times 0.6 + 0.3 \times 0.4} = \frac{0.12}{0.54} \approx 0.2$$

Now we would like to represent G as a linear combination of A, B, and C. That is to find α, β, and γ so that $G = \alpha A + \beta B + \gamma C$, where $0 \le \alpha, \beta, \gamma \le 1$ and $\alpha + \beta + \gamma = 1$. Let $s = C'G$, $t = GC$ in Fig. 3b. By applying the Menelaus's theorem to the triangle $\triangle CAC'$ and the line BB', we have the equation $r(p + p')s = r'p't$. Thus $s = \frac{r}{r'p'}t$ holds because $p + p' = 1$. From this equation, $s + t = \frac{r + r'p'}{r}t$, and thus we have.

$$\frac{s}{s+t} = \frac{r'p'}{r+r'p'}, \qquad \frac{t}{s+t} = \frac{r}{r+r'p'}$$

Since $C' = p'A + pB$,

$$G = \frac{t}{s+t}C' + \frac{s}{s+t}C = \frac{r}{r+r'p'}(p'A+pB) + \frac{r'p'}{r+r'p'}C$$

$$= \frac{1}{r+r'p'}(rp'A+rpB+r'p'C).$$

So we have $\quad \alpha = \dfrac{rp'}{r+r'p'}, \quad \beta = \dfrac{rp}{r+r'p'}, \quad \gamma = \dfrac{r'p'}{r+r'p'}$

From these results we can get the following intended equations.

$$\alpha : \beta = p' : p, \quad \beta : \gamma = rp : r'p' = \frac{rpq}{q} : \frac{r'p'q'}{q'} = q' : q,$$

and $\gamma : \alpha = r' : r$.

As a special case, we suppose the numbers a, b, and c are assigned to the three keys A, B, and C, respectively. Then we have:

$$p' = \frac{a}{a+b}, \quad p = \frac{b}{a+b}, \quad q' = \frac{b}{b+c}, \quad q = \frac{c}{b+c}, \text{ and}$$

$$r' = \frac{c}{c+a}, \quad r = \frac{a}{c+a}$$

Thus the equation

$$p'q'r' = \frac{abc}{(a+b)(b+c)(c+a)} = pqr$$

is satisfied: i.e. the three lines AA', BB', and CC' meet at the same point G from the Ceva's Theorem. Because of

$$r+r'p' = \frac{a}{c+a} + \frac{ca}{(c+a)(a+b)} = \frac{a(a+b+c)}{(a+b)(c+a)},$$

we have:

$$\alpha = \frac{rp'}{r+r'p'} = \frac{a}{c+a}\frac{a}{a+b}\frac{(a+b)(c+a)}{a(a+b+c)} = \frac{a}{a+b+c}$$

Similarly, $\beta = \dfrac{b}{a+b+c}$ and $\gamma = \dfrac{c}{a+b+c}$. And thus we have the expression for G as follows:

$$G = \frac{aA+bB+cC}{a+b+c}$$

3.2 The Methods of Choosing 3 Keys

One of the most important issues in trichotomic approach is how to choose 3 keys. In this section, we discuss possible 3 types of choosing methods.

Type 1 (Choose 3 Keys Manually):
The most orthodox choice method might be to choose them according to the known trichotomic conceptualization. As we have pointed out in Sect. 1, it is popular to capture and represent concepts and objects based on three contrasting points of view. For example, the expression ten–chi–jin (the heaven–the earth–the humans) indicates the three elements of the world, or great nature. This expression is also used for the components for success by turning the meanings into; time/luck–expertise/knowledge–cooperation/harmony of humans. In the next Sect. 3.3, we choose physics, chemistry, and mathematics as the 3 fundamental subjects for natural sciences; which is an example case study in this type of choosing 3 keys (key words).

Type 2 (Choose 2 Keys Manually and Set the Rest Key as Others):
The next possible method of choosing 3 keys is to firstly choose 2 keys according to dichotomic conceptualization, and choose the 3rd one for measuring based on a different point of view. In Sect. 3.4, we choose social sciences, category 300, and natural sciences, category 400, from the top category of NDC system at first as the 2 keys, then we choose other categories (from 000 to 900 avoiding 300 and 400) as a whole for the 3rd key.

Type 3 (Choose 3 Keys by Computation):
The last possible method of choosing 3 keys is to choose them by computation using such as similarity, closeness, distance, etc. values. In Sect. 3.5, we show an example of choosing 3 faculties LA, LT, and SC as the triple of faculties that have minimum sum value of 3 similarity values between two of them among all combinations of 3 faculties.

3.3 Case Study 1: Physics-Chemistry-Mathematics Analysis

As the first case study for trichotomic analysis, we take up physics, chemistry, and mathematics as the three keys and investigate the students' interest areas. We use the NDCs 420, 430, and 410 as the indexes for them because these are for physics, chemistry, and mathematics, respectively. To be more precise, the actual NDC number's range for physics is $420 \leq NDC < 430$, and in the same way for chemistry and mathematics. The weights of a patron are taken by the numbers of borrowed books having the corresponding keys; for example $a =$ the number of borrowed books of NDC number 420, or from 420 up to 420.99 in the records for the patron.

Fig. 4 Scattergram of all students' interest ranges for physics-chemistry-mathematics in trichotomic representation

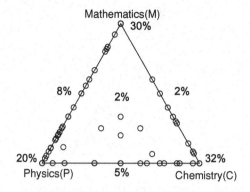

Figure 4 shows the scattergram for all students in trichotomic representation together with the percentages according to borrowing patterns. The borrowing patterns are divided into 7 types; 3 types for those students who borrowed books of only one of the three keys, i.e. physics (P), chemistry (C), and mathematics (M), 3 types for those who borrowed only two categories, i.e. P-C, C-M, or M-P, and the rest 1 category for those who borrowed all 3 categories of books, i.e. P-C-M.

An interesting finding is that only 2 % of students borrowed books from three categories and 83 % from only one category. Different from our prediction, for the ratios for one category, the order is C (32 %) > M (30 %) > P (20 %). However for the pairs, P-M (8 %) > P-C (5 %) > C-M (2 %). From such a seemingly contradictory result, we may tell that chemistry is the most important subject as a single study subject for students as a whole. Thus part of physics' importance comes from that students need to learn physics in order to learn chemistry as it gives theoretical background of chemistry. Similarly, mathematics is important in combination with physics because it gives fundamental concepts and useful tools for analysis for physics.

3.4 Case Study 2: Social Sciences-Natural Sciences-Other Analysis

It is quite popular to divide a person according to his or her major in Japan, whether the major is considered to be science-related or liberal-arts-related. In this section, we take the NDC categories 400 (Natural Sciences, or N) as the representative to science-related subjects and 300 (Social Sciences, or S) as to arts-related subjects. We add up other NDC categories as the third key (Other, or Oth) in order to show the width of interest areas.

We chose the 11 students who appeared in Fig. 1 as the representatives. Figure 5 (left) shows the scattergram for these students in the trichotomic representation.

It is interesting to see that students in the same affiliation are close each other in Fig. 5 (left). Three LA students are located in the left lower area, from vertex S for social sciences to the middle area toward other on the side for S-Oth. Among these

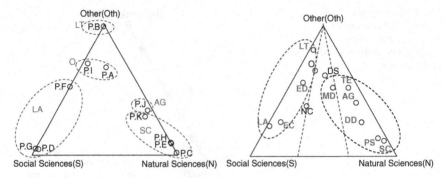

Fig. 5 Scattergram for top 11 students (*left*) and faculties (*right*) in trichotomic representation with social sciences-natural sciences-other as keys

3 students only P.F is separated as this student is different in his or her interest area range.

SC students, on the other hand, are located in the lower area at the right side; from vertex N to around the 2:1 point toward the vertex Oth. Similar to LA case, one student P.K is separated from other students because he or she has wider interest are range.

The student P.J, who is the only student affiliated in AG, is located very close to P.K. Even though they are different in affiliation and patron type, they are very similar in their interest areas.

The 2 students P.A and P.I classified as Other (O) are located in the upper area relatively near the vertex Oth. This result is understandable because the students in O might not strongly related to either social or natural sciences.

The only student P.B who affiliated in Letter (LT) is located to the vertex Oth. According to Fig. 1, he or she is strongly literature-oriented and should locate far away from both social and natural sciences.

Figure 5 (right) shows the scattergram for 12 faculties plus other (O) in the same keys as in Fig. 5 (left). As we have expected the faculties considered as liberal-arts-related are located near the left side and those considered as science-related are located near the vertex N. In the figure are two dashed lines for dividing the whole area into 3 smaller areas; social science oriented, natural science oriented, and intermediate.

It is interesting to see that Medicine (MD) is located in the intermediate area. Even though MD is normally considered as being included in the science-related, it is more liberal-arts oriented than other science-related subjects, probably because they are more expected to be interested in humans not only from biological view point but also from psychological and behavioral view points.

The two faculties DS (Faculty of Design, or Faculty of Arts and Engineering in Japanese) and NC (Faculty of New Century, or Faculty of 21st Century in Japanese) are difficult to specify if they are liberal arts-related or science-related. In Fig. 5 (right), we can see that both of them are located in the intermediate area. As we see further, DS is more science-related than arts-related and NC is located in the left most part of intermediate area and thus we may consider it is almost arts-related.

The others group (O) is also near the border line between arts-related and intermediate areas just like NC. So the students in others group are rather arts-oriented than science-oriented in general.

As we see the heights, or the degree of closeness to the vertex Oth, of faculties, Letter (LT) is the highest, which means that the NDC category 900 (Literature) is included in the other subject and thus the students like P.B are located high in the trichotomic representation.

On the other hand, Sciences (SC) and Pharmaceutical (PS) are strongly centered to natural sciences than any other faculties. Among liberal arts-related faculties, Law (LA) and Economy (EC) are most strongly centered to social sciences.

We can recognize also in this case study that trichotomic representation gives new findings about the characteristic behaviors not only of the 11 students shown in the previous section but also of the faculties as well.

3.5 Case Study 3: Locating Faculties by Choosing 3 Most Separated Faculties as Keys

In this case study we apply the 3rd method (type 3) of choosing keys to understanding the mutual relations of faculties, where we choose 3 faculties as the keys for comparison. In order to arrange the faculties according to the pair-wise similarity values, we will choose the 3 faculties which have the minimum value of sum of the similarities of 3 pairs among all the 3-tuples. We choose this method because we would like to choose 3 faculties so that they are far away each other and thus the faculties are arranged in a wide area. In this way, we will be able to see the differences of faculties' interest areas clearly.

As we can guess from Fig. 5 (right), the 3-tuple consisting of Law (LA), Letter (LT), and Sciences (SC) is chosen as the result. Actually the similarity value between LA and SC is 0.06, which is the minimum value among all the pair-wise similarity values.

The faculty of LT is not so far off from SC as LA. The similarity value between LT and SC is 0.12, which is the 3rd smallest one. The second smallest pair-wise similarity value is 0.10 between LA and PS (Pharmaceutical). PS and SC are very close each other in terms of interest area profile. Their similarity value is 0.99, which is the maximum value of pair-wise similarity values of faculties.

The similarity value between LA and LT is 0.53, and thus the sum value of the 3-tuple of LA-LT-SC is $sim(LA, LT) + sim(LT, SC) + sim(SC, LA) = 0.53 + 0.12 + 0.06 = 0.71$. Incidentally, if we replace SC with PS, the result is $sim(LA, LT) + sim(LT, PS) + sim(PS, LA) = 0.53 + 0.14 + 0.10 = 0.77 > 0.71$.

Fig. 6 Scattergram for faculties with the keys of *LA*, *LT*, and *SC*

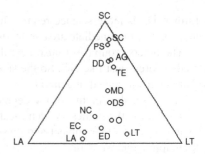

Figure 6 shows the resulting scattergram. In comparison with Fig. 5 (right), social sciences is replaced with the faculty LA, which is the closest to the vertex for social sciences, natural sciences is replaced with SC, closest to the vertex for natural sciences, and others vertex is replaced with LT, which is also the closest faculty to the vertex. Thus the mutual closeness of faculties are basically the same in both figures.

A problem in this figure is the location of the key faculties. Even though we choose LA, LT, and SC as the keys, they are located away from the vertex. For example, LA is located near the bottom line between the vertexes of LA and LT, however it is located not very close to vertex of LA horizontally. As we see the similarity values more precisely, $sim(LA, LA) = 1$ and $sim(LA, LT) = 0.53$, thus their ratio is about 1 : 2. So LA's horizontal location is about 1 : 2 in the bottom line between the vertexes LA and LT. Similarly, LT is located at about 2 : 1 position horizontally.

In order to overcome this problem, we take the ratio of the "distance" values instead of similarity values. We define the distance by: $Dist(A, B) = 1 - Sim(A, B)$. Since $0 \leq Sim \leq 1$, the distance value has the same range as similarity value (i.e. $0 \leq Dist \leq 1$). From this definition, $Dist(LA, LA) = 0$ (and $Dist(LA, LT) = 0.47$), thus LA is located on the vertex for LA. Similarly LT and SC are located on the vertexes for LT and SC, respectively.

Figure 7 (left) shows the resulting triangular scattergram using the distance ratios. We can see the mutual closeness more clearly than Fig. 6 in this figure. In the figure, liberal-arts-related faculties are located almost in the bottom area and are stretching from left to right.

The O (Other) is also located in this area, which indicates that the students classified as other are basically liberal-arts-oriented and thus have little interest in natural sciences. The faculty of NC (New Century, or the 21st Century Program) is also located in this area. Even though NC was founded for the students who are interested in both liberal-arts-related subjects and natural sciences subjects, they are basically liberal-arts-oriented rather than natural sciences oriented.

Among these liberal-arts-oriented faculties, those located close to the LA vertex are rather concentrated in social sciences and those located close to LT have wider interest area. In this respect, the students of LT and O have wider interest areas than other faculty students.

The rest faculties from Sciences (SC) to Design, or Arts and Engineering (DS) are almost located on a line, which indicates that interest area of these natural sciences oriented faculty students are balanced in terms of liberal-arts-oriented subjects.

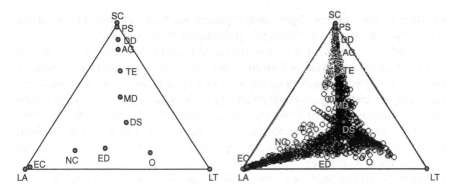

Fig. 7 Scattergram for faculties (*left*) and all students (*right*) with the Keys of *LA*, *LT*, and *SC*, Using distance values instead of similarity values

Among these 7 faculties, DS and Medical (MD) students are more liberal-arts-oriented than others. Considering that DS intends to collect students from both field of arts and engineering, this result might be natural and reasonable.

On the other hand, it is different from our prediction that MD is located near the center of the triangle because it looks like a typical natural sciences related faculty. However as we have this result and think about it character, it may be a natural result: In order to become a good doctor, the students not only supposed to study natural sciences related subjects like mathematics, physics, chemistry, etc. but also to study wider subjects such as biology, psychology, etc. Maybe it is preferable for them to read even wider areas including literatures, novels, stories, etc. in order to be able to understand what humans are. From this point of view it may be a reasonable result that MD located in this position.

Figure 7 (right) shows the scattergram for all students. It is interesting that some students are located nearly on the vertexes for LA and SC, there are no students on or near the vertex for LT. This result indicates that there are not students who we can call typical LT students. Their interest area profile patterns vary a lot and the profile for LT is just the result as a whole.

4 Concluding Remarks

Firstly in this chapter, we explained our idea of trichotomic approach to concept representation, especially when the concept is difficult to specify directly in ordinary words. Trichotomic representations for special purposes have investigated by many scholars. For example, Peirce [8] dealt with three principles for conception; first, second, and third. Widmeyer [9] applied to the design of information systems. Aebischer [1] discuss the relationship between ICT and energy using Spreng's triangle. Our idea is to represent such concept by specifying the relative distances to 3 keys (e.g. words, subjects, objects such as patrons and books, etc.) in an approximate fashion; which we call a trichotomic representation of the concept.

Secondly, we presented three case studies of trichotomic representations of students and faculties of Kyushu University in order to demonstrate the usefulness of trichotomic approach based on the library's loan records. In the first case study, we chose physics, chemistry, and mathematics because these are three fundamental subjects for students affiliated in the (natural) sciences-related faculties. We observed how the 11 representative students are located and discussed what we can find from the trichotomic representations. We also applied this method to faculties and observed their characteristic features.

In the second case study, we chose the NDC numbers 300 (social sciences) as to represent the liberal arts-related subjects and 400 (natural sciences) as for the (natural) sciences-related subjects, and analyzed the scattergrams for the top 11 students and for faculties just like the first case study. From these case studies we were able to confirm that trichotomic representation is another good tool for analysis in order to obtain new findings.

In the third case study, we chose three faculties by a given criterion and investigate the relative locations of faculties in terms of these three faculties as keys. We found that use of distance measure is more convenient than use of similarity measure in this case.

What we have to do toward the future include (1) to study the representation methods in a triangular form more generally when the ratios do not satisfy the condition $pqr = p'q'r'$, where such cases will appear when humans give them based on their institution, (2) to develop a system with a good interface so that we can carry out more case studies and experiments with ease, and (3) to automatize the analysis methods so that we can analyze various types data (semi-)automatically.

Acknowledgments This work was supported in part by the Ministry of Education, Science, Sports and Culture, Grant-in-Aid for Scientific Research (C), 24500318, 2013.

References

1. B. Aebischer, ICT and energy: methodological issues and Spreng's triangle. (The European e-Business Report 2008), pp. 265–269, http://ec.europa.eu/enterprise/archives/e-business-watch/key_reports/documents/EBR08.pdf
2. T. Minami, H. Sazuka, S. Hirokawa, T. Ohtani, in *Living with ZK - An Approach towards Communication with Analogue Messages*, ed. by L.C. Jain, R.K. Jain. 1998 Second International Conference on Knowledge-Based Intelligent Electronic Systems (KES 1998), pp. 369–374 (1998)
3. T. Minami, Book Profiling from Circulation Records for Library Marketing—Beginning from Manual Analysis toward Systematization—. *International Conference on Applied and Theoretical Information Systems Research (ATISR 2012)*, pp. 15 (2012)
4. T. Minami, K. Baba, in Investigation of Interest Range and Earnestness of Library Patrons from Circulation Records. *International Conference on e-Services and Knowledge Management (ESKM 2012), as a part of the 1st IIAI International Conference on Advanced Applied Informatics (IIAI-AAI 2012), IEEE CPS*, pp. 25–29, (2012). doi:10.1109/IIAI-AAI2012.15

5. T. Minami, in Interest Area Analysis of Person and Group Using Library's Circulation Records. *Proceedings of IADIS International Conference Information Systems (IS 2013)*, p. 8 (2013)
6. T. Minami, S. Hirokawa, K. Baba, E. Amano, in A Trichotomic Approach to Approximate Representation of Concepts—With its Application to Library Data Mining—. *International Conference on Advanced Software Engineering and Information Systems (ICASEIS) in IIAI AIT 2013*, p. 6 (2013)
7. NHK, Close-Up Today: "Pamyu-Pamyu", "Je-Je-Je"— Mystery of Great Propagation of Onomatopoeia— (2013). http://www.nhk.or.jp/gendai/kiroku/detail02_3362_all.html (in Japanese)
8. C.S. Peirce, The architecture of theories. The Monist. **1**(2), 161–176 (1891), http://www.jstor.org/stable/27896847
9. G.R. Widmeyer, The trichotomy of processes: a philosophical basis for information systems. Aust. J. Inf. Syst. **11**(1), 3–11 (2003)

Hard Optimization Problems in Learning Tree Contraction Patterns

Yasuhiro Okamoto and Takayoshi Shoudai

Abstract A *tree contraction pattern* (*TC-pattern*) is an unordered tree-structured pattern common to given unordered trees, which is obtained by merging every uncommon connected substructure into one vertex by edge contraction. In order to extract meaningful and hidden knowledge from tree structured documents, we consider a minimal language (MINL) problem for TC-patterns. The MINL problem for TC-patterns is to find a TC-pattern t such that the language generated by t is minimal among languages, generated by TC-patterns, which contain all given unordered trees. Recently, [8] showed that the MINL problem for TC-patterns is computable in polynomial time if there are infinitely many vertex labels. In this chapter, we discuss two optimization versions of the MINL problem, which are called **MINL with Tree-size Maximization (MAX MINL)** and **MINL with Variable-size Minimization (MIN-MAX MINL)**. We show that **MAX MINL** is NP-complete and **MIN-MAX MINL** is MAX SNP-hard.

Keywords Graph mining · MAXSNP-hard · NP-hard · Optimization problems · Tree contraction pattern · Tree-structured pattern

1 Introduction

Many documents such as Web documents or XML files have tree structures. In order to extract meaningful and hidden knowledge from such documents, we need tree-structured patterns that can explain them. As unordered tree-structured patterns, a tree pattern [1, 2], a type of object [4], a tree-expression pattern [7], and an unordered

Y. Okamoto · T. Shoudai (✉)
Department of Informatics, Kyushu University, Fukuoka 819-0395, Japan
e-mail: shoudai@inf.kyushu-u.ac.jp

Y. Okamoto
e-mail: yasuhiro.okamoto@inf.kyushu-u.ac.jp

R. Y. Lee (ed.), *Applied Computing and Information Technology*,
Studies in Computational Intelligence 553, DOI: 10.1007/978-3-319-05717-0_6,
© Springer International Publishing Switzerland 2014

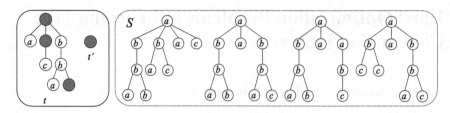

Fig. 1 In this chapter, a contractible vertex of a TC-pattern is represented by *dark-colored circles*. Both TC-patterns t and t' are matched by all unordered trees in S, but t' is an overgeneralized TC-pattern for S

term tree [3, 6] have been proposed. In [8], we introduced edge contraction-based tree-structured patterns, called *tree contraction patterns*, as graph patterns that are learnable in polynomial time.

A *tree contraction pattern (TC-pattern)* is a triplet $t = (V_t, E_t, U_t)$ where (V_t, E_t) is an unordered tree with a specified root r_t and U_t is a subset of V_t. A vertex in U_t is called a *contractible vertex*. An unordered tree $T = (V_T, E_T)$ with root r_T matches a TC-pattern t with root r_t if there exists a partition $\mathcal{W} = \{W(u) \mid u \in V_t\}$ of V_T such that (i) for $u \in V_t \setminus U_t$, $W(u)$ includes exactly one vertex, (ii) for any $u \in V_t$, any pair of vertices in $W(u)$ is connected, (iii) $W(r_t)$ includes r_T, and (iv) the tree obtained from T by merging all vertices in $W(u)$ into one vertex for each $u \in U_t$ is isomorphic to (V_t, E_t). In [8], we showed that the pattern matching problem for TC-patterns is computable in polynomial time if all contractible vertices have a constant number of children.

The concept represented by a TC-pattern t, which is called the TC-pattern language of t, is the set of all unordered trees that match t. The minimal language (MINL) problem is the problem of finding a TC-pattern whose TC-pattern language is minimal among TC-pattern languages containing all given unordered trees.

In Fig. 1, we give a set of unordered trees S as a given data. Overgeneralized patterns are meaningless. Then the purpose of this work is to find one of the least generalized TC-patterns that are matched by a given data. In Fig. 1, we give examples of TC-patterns t and t' that are matched by all unordered trees in S. The TC-pattern t' is an overgeneralized TC-pattern, which is meaningless. On the other hand, the TC-pattern t is one of the least generalized TC-patterns that is matched by all unordered trees in S. In Fig. 2, All least generalized TC-patterns for a given data S are given in the right figure. In [8], we showed that the problem for finding a least generalized TC-patterns for a given set of unordered trees is computable in polynomial time. In this chapter, we call such a problem an MINL problem.

MAX SNP is a complexity class introduced by Papadimitriou and Yannakakis [5] with a concept of reduction, called L-reduction, that preserves approximation schemes. An L-reduction is defined as follows. Let Π and Π' be two optimization problems. We say that Π *L-reduces* to Π' if there are two polynomial-time algorithms f, g, and constants $\alpha, \beta > 0$ such that for each instance I of Π:

1. $OPT(f(I)) \leq \alpha \cdot OPT(I)$.

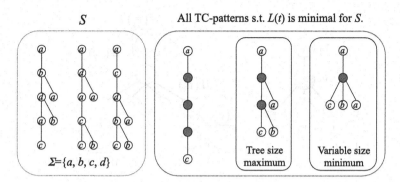

Fig. 2 The least generalized TC-patterns that is matched by all unordered trees in S

2. Given any solution of I' with cost s_2, algorithm g produces a solution of I with cost s_1 such that $|OPT(I) - s_1| \le \beta \cdot |(OPT(f(I)) - s_2)|$.

Papadimitriou and Yannakakis showed that if Π L-reduces to Π' and there is a polynomial-time algorithm for Π' with worst case error ϵ, then there is a polynomial-time approximation algorithm for Π with worst-case error $\alpha\beta\epsilon$. If Π' has a polynomial time approximation scheme (PTAS), then so does Π. A problem is **MAX SNP**-*hard* if every problem in MAX SNP can be L-reduced to it. Since the composition of two L-reductions is also an Ł-reduction, a problem is **MAX SNP**-hard if a **MAX SNP**-hard problem can be L-reduced to it. It is known that no MAX SNP-hard problem has a polynomial time approximation scheme, unless NP = P. Therefore, it is very unlikely for a MAX SNP-hard problem to have a PTAS.

In this chapter, we consider the following two problems. Firstly, **MINL with Tree-size Maximization (MAX MINL)** is the problem of finding a TC-pattern t such that the TC-pattern language of t is minimal and the number of vertices in t is maximum. Secondly, **MAX MINL with Variable-size Minimization (MIN-MAX MINL)** is the problem of finding a TC-pattern t such that the TC-pattern language of t is minimal, the number of vertices in t is at least a given positive integer K, and the number of contractible vertices in t is minimum. Then we prove that **MAX MINL** is NP-complete and **MIN-MAX MINL** is MAX SNP-hard. These results show the hardness of finding the optimum TC-pattern representing a given data.

2 Tree Contraction Patterns

For a tree T, we denote by $V(T)$ the set of vertices of T and by $E(T)$ the set of edges. For $V' \subseteq V(T)$, we denote by $T[V']$ the subtree of T induced by V'.

Definition 1 Let T and S be trees. An S-*witness structure* of T is a partition $\mathcal{W} = \{W(u) \subseteq V(T) \mid u \in V(S)\}$ of $V(T)$ satisfying the following conditions.

1. For any $u \in V(S)$, $T[W(u)]$ is connected, and

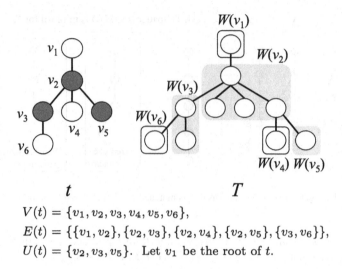

$$V(t) = \{v_1, v_2, v_3, v_4, v_5, v_6\},$$
$$E(t) = \{\{v_1, v_2\}, \{v_2, v_3\}, \{v_2, v_4\}, \{v_2, v_5\}, \{v_3, v_6\}\},$$
$$U(t) = \{v_2, v_3, v_5\}. \text{ Let } v_1 \text{ be the root of } t.$$

Fig. 3 In this chapter, a contractible vertex of a d-TC-pattern is represented by *dark-colored circles*. Then, $\mathcal{W} = \{W(v_1), \ldots, W(v_6)\}$ is a t-witness structure of T

2. for any $u, u' \in V(S)$ $(u \neq u')$, $\{u, u'\} \in E(S)$ if and only if there is an edge $\{v, v'\} \in E(T)$ such that $v \in W(u)$ and $v' \in W(u')$

We call each set $W(u) \in \mathcal{W}$ the *S-witness set* of u. When T has an S-witness structure, T can be transformed to S by contracting each of the S-witness sets into one vertex by edge contractions. A *tree contraction pattern* t (*TC-pattern*, for short) is a triplet $t = (V_t, E_t, U_t)$ where (V_t, E_t) is an unordered tree and U_t is a subset of V_t. We call an element of U_t a *contractible vertex*. In particular, if any contractible vertex of t has at most d children for a constant positive integer d, we call it a d-TC-pattern.

Let Σ be a finite alphabet. Each vertex of a TC-pattern t has its vertex label in Σ. For a d-TC-pattern t, we denote by $V(t)$ the set of vertices of t, by $E(t)$ the set of edges of t, and by $U(t)$ the set of contractible vertices of t.

Let t be a d-TC-pattern with vertex labels in Σ. We say that a d-TC-pattern t with root r_t *matches* an unordered tree T with root r_T if there is a t-witness structure $\mathcal{W} = \{W(u) \subseteq V(T) \mid u \in V(t)\}$ of T such that (i) for all $v \in V(t) \setminus U(t)$, $W(v)$ contains exactly one vertex $u \in V(T)$ of the same label as v, and (ii) $r_T \in W(r_t)$. For example, we show in Fig. 3 a d-TC-pattern t and an unordered tree T that has a t-witness structures $\mathcal{W} = \{W(v_1), \ldots, W(v_6)\}$ of T. Each witness set in $\mathcal{W}$ is specified with a squared area. Thus, T matches t. We formally define the pattern matching problem for d-TC-patterns as follows:

d-TC-**PATTERN MATCHING**

Instance: A d-TC-pattern t and an unordered tree T.
Question: Does T match t?

In [8], we proved the following theorem:

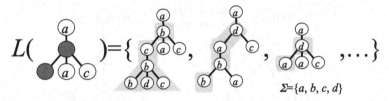

Fig. 4 An example of d-TC-pattern languages over $\Sigma = \{a, b, c, d\}$

Theorem 1 *Let d be a constant positive integer.*

1. *d-TC-PATTERN MATCHING is NP-complete even if the maximum degree of a given unordered tree is equal to 3.*
2. *d-TC-PATTERN MATCHING for a given d-TC-pattern t and a given tree T is solved in $O(nN(n + \sqrt{N}))$ time, where $n = |V(t)|$ and $N = |V(T)|$.*

3 A Polynomial-Time Algorithm for Finding Minimal d-TC-Pattern Languages

TC_Σ^d and $\mathcal{T}_\Sigma$ denote the set of all d-TC-patterns and the set of all unordered trees with vertex labels in Σ, respectively. For a d-TC-pattern $t \in TC_\Sigma^d$, we define the d-TC-pattern language of t as $L(t) = \{T \in \mathcal{T}_\Sigma \mid T$ matches $t\}$. In Fig. 4, we give an example of d-TC-pattern languages over $\Sigma = \{a, b, c, d\}$. A d-TC-pattern language $L(t)$ is *minimal* for (S, TC_Σ^d) if (1) $S \subseteq L(t)$ and (2) $L(t') \subsetneq L(t)$ implies $S \not\subseteq L(t')$ for any $t' \in TC_\Sigma^d$. For $S \subset \mathcal{T}_\Sigma$, a d-TC-pattern t is called a *least generalized d-TC-pattern* for S if $L(t)$ is minimal for (S, TC_Σ^d).

Minimal Language (MINL) Problem for TC_Σ^d

Instance: A nonempty finite subset S of $\mathcal{T}_\Sigma$.

Question: Find a least generalized d-TC-pattern $t \in TC_\Sigma^d$ for S.

In [8], we proposed two groups of refinement operators for d-TC-patterns, which are shown in Fig. 5. Those refinement operators can be used to extend a d-TC-pattern as much as possible while $S \subseteq L(t)$ holds. We start with a d-TC-pattern t consisting of only one contractible vertex, and try to replace every contractible vertex v in t with one of the structures on the right-hand sides of the arrows in Fig. 5 if it is possible. Below, we show an outline of our polynomial-time algorithm for computing **MINL Problem for TC_Σ^d**.

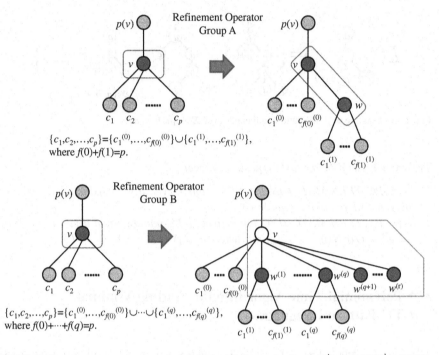

Fig. 5 Refinement operator groups A and B: A and B contain the $O(2^d)$ and $O(N_{\min}^d |\Sigma(S)|)$ refinement operators, respectively, where $N_{\min} = \min_{T \in S} |V(T)|$ and $\Sigma(S) = \{\sigma \in \Sigma \mid \sigma$ appears in all trees in $S\}$

Algorithm MINL;

Input: A nonempty finite set S of unordered trees.

Step 1: Let v be a new contractible vertex, and

$\quad\quad t := (\{v\}, \emptyset, \{v\})$.

Step 2: Until a d-TC-pattern t cannot be updated anymore,

do the next steps:

$\quad$ 2-1: Choose a contractible vertex v from $U(t)$ and

$\quad\quad$ a refinement operator σ in Fig. 5.

$\quad$ 2-2: Construct a new d-TC-pattern t' from t by

$\quad\quad$ applying σ to v.

$\quad$ 2-3: If all $T \in S$ matches t', then $t := t'$.

In [8], we proved the following theorem:

Theorem 2 *We assume that there are infinitely many vertex labels in Σ.* **MINL Problem** *for TC_Σ^d for a nonempty finite subset S of T_Σ is computed in $O(N_{\min}^{d+3} N_{\max} (N_{\min} + \sqrt{N_{\max}})|S|)$ time, where $N_{\min} = \min_{T \in S} |V(T)|$, $N_{\max} = \max_{T \in S} |V(T)|$.*

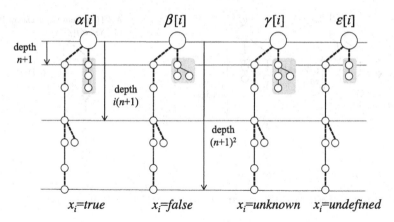

Fig. 6 Four subtrees for representing that $x_i = true, false, unknown$, and $undefined$: A *broken line* specifies a path of length $n + 1$ from its *upper endpoint* to its *lower endpoint*

4 Minimal Language Optimization Problems

In this section, we discuss MINL problems with optimizing the size of an output d-TC-pattern. First, we show the following special MINL problem is hard to compute.

MINL with Tree-size Maximization (MAX MINL)
Instance: A nonempty finite subset S of $\mathcal{T}_\Sigma$ and a positive integer K.
Question: Is there a least generalized d-TC-pattern t for S that satisfies $|V(t)| \geq K$?

Theorem 3 **MAX MINL** *is NP-complete even if $d = 2$ and $|\Sigma| = 1$.*

Proof Membership in NP is obvious. We transform 3-SAT to this problem. Let $U = \{x_1, \ldots, x_n\}$ be a set of variables and $C = \{c_1, \ldots, c_m\}$ be a collection of clauses over U such that each clause c_j $(1 \leq j \leq m)$ has $|c_j| = 3$. We use only *blank* for vertex labels, i.e., $\Sigma = \{blank\}$. Firstly, for each variable x_i $(1 \leq i \leq n)$, we make four trees for representing that $x_i = true, false, unknown$, and $undefined$ (Fig. 6). Let $I_n = n^2 + 4n + 6$. I_n shows the number of vertices of $\alpha[i]$ and $\beta[i]$ for all i $(1 \leq i \leq n)$. The numbers of vertices of $\gamma[i]$ and $\epsilon[i]$ are $I_n + 1$ and $I_n - 1$, respectively. We show the next two claims:

Claim 1 For the trees $\alpha[i]$ and $\beta[i]$ $(1 \leq i \leq n)$, the number of vertices of any d-TC-pattern that is matched by the two trees is at most $I_n - 1$.

Claim 2 Let $\chi[i]$ $(1 \leq i \leq n)$ be one of $\alpha[i], \beta[i], \gamma[i]$, and $\epsilon[i]$ in Fig. 6. For the trees $\chi[i]$ and $\chi[j]$ $(i \neq j)$, the number of vertices of any d-TC-pattern that is matched by the two trees is at most $I_n - (n + 1)$.

For example, in the case of $\alpha[i]$ and $\alpha[j]$ $(i \neq j)$, any least generalized d-TC-pattern for the trees has at most $I_n - (n + 1)$ vertices (Fig. 7).

We construct trees $T_1, \ldots, T_m$ from $c_1, \ldots, c_m$. We use the trees $\alpha[i], \beta[i]$, and $\gamma[i]$ $(1 \leq i \leq n)$ in Fig. 6 as subtrees. The root of T_j has exactly 7 children. Each

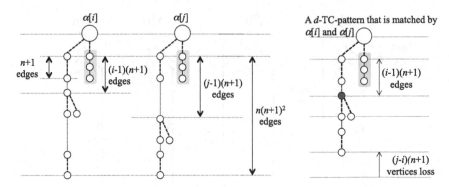

Fig. 7 Any least generalized d-TC-pattern for $\alpha[i]$ and $\alpha[j]$ loses at least $n + 1$ vertices if $i \neq j$

subtree rooted at each child shows a truth assignment which satisfies the clause c_j. For example, let $c_j = \{x_1, \bar{x}_2, x_3\}$. In Fig. 8, the colored area shows one of the 7 assignments that satisfies c_j. Moreover we construct one special tree T (Fig. 9). Then $|V(T)| = 7(n + 1)^2 + 7nI_n - 5n + 1$. Let $S = \{T_1, \ldots, T_m, T\}$ be an input data. Lastly we set $K = |V(T)| - n$. We have the next claim:

Claim 3 There is a least generalized d-TC-pattern t for S that satisfies $|V(t)| = K$ if and only if 3-SAT has a truth assignment that satisfies all clauses in C.

Proof of Claim 3 Without loss of generality, we assume that $n > d$. We suppose that there is a truth assignment which satisfies all clauses $c_1, \ldots, c_m$. By observing the truth assignment, we design a d-TC-pattern with K vertices. For example, for $(x_1, x_2, \ldots, x_i, \ldots, x_n) = (true, false, \ldots, true, \ldots, false)$, we make a d-TC-pattern shown in Fig. 10. We can easily see that a d-TC-pattern constructed in this way has exactly K vertices and is matched by all unordered trees in S. Conversely, we suppose that there is a d-TC-pattern $t = (V(t), E(t), U(t))$ with $|V(t)| \geq K$. Since the special tree T (Fig. 9) has $K + n$ vertices, t is obtained from T by at most n edge contractions. The vertices at depth 1 and 2 of T are not corresponding to any contractible vertex of t, since all the vertices have more than d children. If there is a vertex at depth 1 or 2 that is corresponding to a contractible vertex of t, t must be obtained from T by more than $n^2 + n + 1$ edge contractions. In addition, the root of t is not a contractible vertex. Therefore, edge contractions on T must be applied to edges whose lower endpoint locates at depth more than 3. Since t is matched by all T_j ($1 \leq j \leq m$) and T, from Claims 1 and 2, each of n edge contractions on T must be applied to each n subtree rooted at depth 2. Moreover the edge contraction must be applied to the rightmost subtree of size 4 rooted at 3. Therefore t must be of the form specified in Fig. 10. Finally we have a truth assignment from the rightmost subtree specified with a colored area. (*End of Proof of Claim 3*)   □

MAX MINL with Variable-size Minimization (MIN-MAX MINL)
Instance: A nonempty finite subset S of $\mathcal{T}_{\Sigma}$ and a positive integer K.

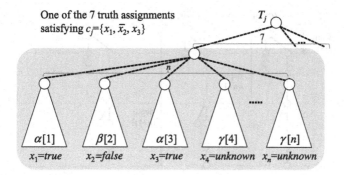

Fig. 8 A subtree for representing one of the truth assignments for $c_j = \{x_1, \bar{x}_2, x_3\}$ by using two trees given in Fig. 6. The *double lines* specify the paths of length n

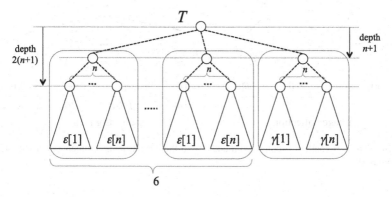

Fig. 9 A special sample tree

Problem: Find a least generalized d-TC-pattern t for S that satisfies the smallest number of contractible vertices in t while $|V(t)| \geq K$.

Theorem 4 **MIN-MAX MINL** *is MAX SNP-hard even if $d = 2$ and $|\Sigma| = 4$.*

Proof We assume that I is an instance of **MAX 2-SAT**. The instance I consists of a set of variables $U = \{x_1, \ldots, x_n\}$ and a set of clauses $C = \{c_1, \ldots, c_m\}$, where $|c_j| = 2$ for all j $(1 \leq j \leq m)$. Let $\Sigma = \{\alpha, \beta_1, \beta_2, blank\}$.

For any variable x_i $(1 \leq i \leq n)$, we construct a tree $T(x_i)$, the top tree in Fig. 11, in order to represents the value of each clause c_j $(1 \leq j \leq m)$ according to x_i. As small components, we use five trees $\alpha[j]$, $\beta_1[j]$, $\beta_2[j]$, $\gamma[j]$, and $\epsilon[j]$ for a clause c_j. A broken line in Figs. 11, 12, 13, 14 and 15 is a path of length $m + 1$ between its upper endpoint and lower endpoint. The root of $T(x_i)$ has two subtrees (specified with *positive* and *negative*. The positive and negative subtrees represent the value of each clause in C when $x_i = true$ and $x_i = false$, respectively. The positive subtree (resp. negative subtree) consists of the m small subtree $\chi[1], \ldots, \chi[m]$ (resp. $\chi'[1], \ldots, \chi'[m]$). Each $\chi[j]$ $(1 \leq j \leq m)$ is defined as follows:

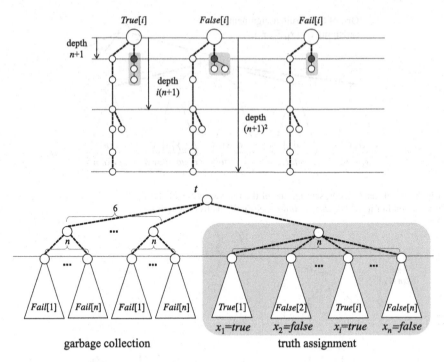

Fig. 10 This d-TC-pattern is outputted if and only if there is a truth assignment which satisfies C (Theorem 3)

$$\chi[j] = \begin{cases} \alpha[j] & \text{if } x_i \in c_j, \\ \beta_1[j] & \text{if } \bar{x}_i \in c_j \text{ and for } i < k, \text{ either } x_k \in c_j \text{ or } \bar{x}_k \in c_j, \\ \beta_2[j] & \text{if } \bar{x}_i \in c_j \text{ and for } k < i, \text{ either } x_k \in c_j \text{ or } \bar{x}_k \in c_j, \\ \gamma[j] & otherwise. \end{cases}$$

In the same way, we define each subtree $\chi'[j]$ $(1 \le j \le m)$.

$$\chi'[j] = \begin{cases} \alpha[j] & \text{if } \bar{x}_i \in c_j, \\ \beta_1[j] & \text{if } x_i \in c_j \text{ and for } i < k, \text{ either } x_k \in c_j \text{ or } \bar{x}_k \in c_j, \\ \beta_2[j] & \text{if } x_i \in c_j \text{ and for } k < i, \text{ either } x_k \in c_j \text{ or } \bar{x}_k \in c_j, \\ \gamma[j] & otherwise. \end{cases}$$

For example, in Fig. 12, we give a tree $T(x_3)$ for an instance $C = \{c_1, c_2, c_3, c_4\}$. Moreover, we make a special tree T_m which has two subtrees. One of the subtrees consists of m trees $\gamma[1], \ldots, \gamma[m]$ and the other consists of m trees $\epsilon[1], \ldots, \epsilon[m]$. Let $S = \{T(x_1), \ldots, T(x_n), T_m\}$. Then, we have the two following claims:

Claim 1 If c_j is satisfied, the subtree for c_j of any least generalized d-TC-pattern t with $|V(t)| \ge K$ has exactly one contractible vertex. $Sat[i]$ in Fig. 15 is such a d-TC-pattern.

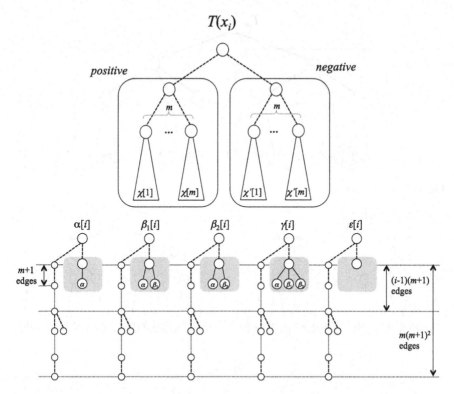

Fig. 11 Tree expression for x_i and five subtrees for a clause c_j. A *broken line* specifies a path of length n from its *upper endpoint* to its *lower endpoint*

$$
\overset{c_1}{(x_1 \vee x_2)} \wedge \overset{c_2}{(\overline{x}_1 \vee x_3)} \wedge \overset{c_3}{(x_2 \vee x_4)} \wedge \overset{c_4}{(\overline{x}_3 \vee x_4)}
$$

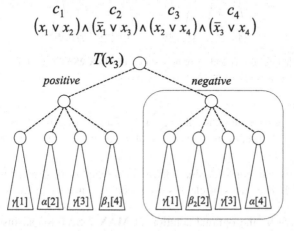

Fig. 12 An example tree for x_3 of an instance of **MAX 2SAT** $C = \{c_1, c_2, c_3, c_4\}$, where $c_1 = \{x_1, x_2\}$, $c_2 = \{\overline{x}_1, x_3\}$, $c_3 = \{x_2, x_4\}$, $c_4 = \{\overline{x}_3, x_4\}$

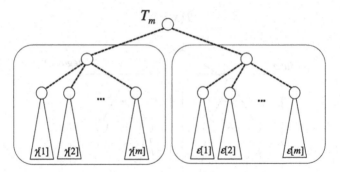

Fig. 13 A special tree T_m

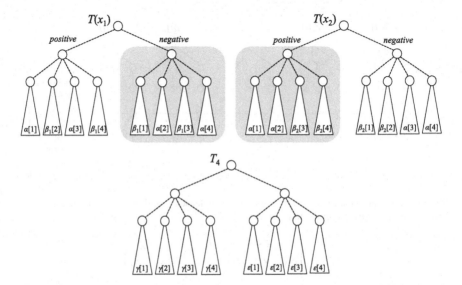

Fig. 14 Example trees for x_1 and x_2 of an instance of **MAX 2SAT** $C = \{c_1, c_2, c_3, c_4\}$, where $c_1 = \{x_1, x_2\}, c_2 = \{\bar{x}_1, x_2\}, c_3 = \{x_1, \bar{x}_2\}, c_4 = \{\bar{x}_1, \bar{x}_2\}$

Claim 2 If c_j is unsatisfied, the subtree for c_j of any least generalized d-TC-pattern t with $|V(t)| \geq K$ has at least two contractible vertex. $Unsat[i]$ in Fig. 15 is such a d-TC-pattern.

Therefore we can conclude that a least generalized d-TC-pattern for S is of the form in Fig. 15. The states of clauses are specified with one of the subtrees in the d-TC-pattern.

Let $OPT(I)$ be the optimal solution of **MAX 2-SAT** for an instance I. It is easy to see that the number of contractible vertices in this d-TC-pattern is $m + (m - OPT(I)) = 2m - OPT(I)$. We show that this transformation $f : $ **MAX 2-SAT** $\rightarrow$ **MIN-MAX MINL** is an L-reduction. We have to show the two following inequalities.

Fig. 15 This d-TC-pattern is outputted if and only if there is a truth assignment which satisfies three clauses in C (Theorem 4)

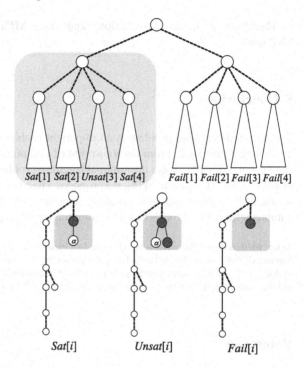

$Sat[1]$ $Sat[2]$ $Unsat[3]$ $Sat[4]$ $Fail[1]$ $Fail[2]$ $Fail[3]$ $Fail[4]$

$Sat[i]$ $Unsat[i]$ $Fail[i]$

1. $OPT(f(I)) \leq \alpha \cdot OPT(I)$ for a constant $\alpha > 0$.
   Actually $OPT(f(I)) = 2m - OPT(I) \leq 4 \cdot OPT(I) - OPT(I) = 3 \cdot OPT(I)$, because $OPT(I) \geq \frac{m}{2}$ (it is always possible to satisfy at least half of the inequality above).

2. For every **MIN-MAX MINL** for $f(I)$ of size m_2, we can, in polynomial time, find a solution of I with m_1 clauses satisfied and

$$|OPT(I) - s_1| \leq \beta \cdot |(OPT(f(I)) - s_2)|$$

   where $\beta = 1$.
   Given d-TC-pattern t for S which satisfies $|V(t)| \geq K$, we can construct, in polynomial time, a truth assignment of C as follows: if the subtree of t which represents $c_i \in C$ is $Sat[i]$ (resp. $Unsat[i]$), c_i is satisfied (resp unsatisfied). We call this algorithm g. Given a feasible solution for $f(I)$ of size s_2, we can, in polynomial time, find a truth assignment for I of size s_1 by using g and

$$s_2 - OPT(f(I)) = (2m - s_1) - (2m - OPT(I))$$
$$= OPT(I) - s_1$$
$$|OPT(I) - s_1| = |OPT(f(I)) - s_2|$$

Therefore f is an L-reduction, and thus **MIN-MAX MINL** is MAX
SNP-hard.   □

5 Conclusions

The minimal language problem is an important problem in learning methods such
as inductive inference. From the viewpoint of computational complexity, we have
shown that it is hard to solve the minimal language problems with optimizing the size
of an output d-TC-pattern. In [8], we proposed a fixed-parameter tractable algorithm
for the d-TC-pattern matching problem. As a next step, we are now studying fixed-
parameter tractable algorithms for the minimal language problem for d-TC-patterns.

Acknowledgments This work was supported by Grant-in-Aid for Scientific Research (C) (Grant
Numbers 23500182, 24500178) from Japan Society for the Promotion of Science (JSPS), and Grant-
in-Aid for Scientific Research on Innovative Areas (Grant Number 24106010) from the Ministry
of Education, Culture, Sports, Science and Technology (MEXT), Japan.

References

1. T.R. Amoth, P. Cull, P. Tadepalli, On exact learning of unordered tree patterns. Mach. Learn.
   44(3), 211–243 (2001)
2. S. Goldman, S. Kwek, in On learning unions of pattern languages and tree patterns. *Proceedings
   of ALT-99, Springer, LNAI 1720*, 1720:347–363 (1999)
3. T. Miyahara, T. Shoudai, T. Uchida, K. Takahashi, H. Ueda, in Polynomial time matching
   algorithms for tree-like structured patterns in knowledge discovery. *Proceedings of PAKDD-
   2000, Springer, LNAI 1805*, pp. 5–16 (2000)
4. S. Nestorov, S. Abiteboul, R. Motwani. Extracting schema from semistructured data. *Proceedings of ACM SIGMOD International Conference on Management of Data*, pp. 295–306 (1998)
5. C.H. Papadimitriou, M. Yannakakis, Optimization, approximation, and complexity classes. J.
   Comput. Syst. Sci. **43**, 425–440 (1991)
6. T. Shoudai, T. Uchida, T. Miyahara, in Polynomial time algorithms for finding unordered tree
   patterns with internal variables. *Proceedings of FCT-2001, Springer, LNCS 2138*, pp. 335–346
   (2001)
7. K. Wang, H. Liu, Discovering structural association of semistructured data. IEEE Trans. Knowl.
   Data Eng. **12**, 353–371 (2000)
8. Y. Yoshimura , T. Shoudai, in Learning unordered tree contraction patterns in polynomial time.
   Proceedings of ILP-2013, Springer, LNAI 7842, pp. 257–272 (2013)

A Web Page Segmentation Approach Using Seam Degree and Content Similarity

Jun Zeng, Brendan Flanagan, Qingyu Xiong, Junhao Wen
and Sachio Hirokawa

Abstract Page segmentation has received great attention in recent years. However, most research has been based on some pre-defined heuristics or visual cues which may be not suitable for large-scale page segmentation. In this chapter, we proposed two parameters: seam degree and content similarity, to indicate the coherent degree of a page block. Instead of analyzing pre-defined heuristics or visual cues, our method utilizes the visual and content features to determine whether a page block should be divided into smaller blocks. We also proposed a principled page segmentation method using these two parameters. An experiment was conducted to determine the relationship between the two parameters and the number of segment results. The empirical results also show that our segmentation method can effectively segment a page into different semantic parts.

Keywords Page segmentation · Seam degree · Content similarity · Semantic segment

J. Zeng (✉) · Q. Xiong · J. Wen
Graduate School of Software Engineering, Chongqing University, Chongqing, China
e-mail: zengjun@cqu.edu.cn

Q. Xiong
e-mail: xiong03@cqu.edu.cn

J. Wen
e-mail: jhwen@cqu.edu.cn

B. Flanagan
Graduate School of Information Science and Electrical Engineering, Kyushu University,
Fukuoka, Japan
e-mail: bflanagan.kyudai@gmail.com

S. Hirokawa
Research Institute for Information Technology, Kyushu University, Fukuoka, Japan
e-mail: hirokawa@cc.kyushu-u.ac.jp

R. Y. Lee (ed.), *Applied Computing and Information Technology*,
Studies in Computational Intelligence 553, DOI: 10.1007/978-3-319-05717-0_7,
© Springer International Publishing Switzerland 2014

1 Introduction

Web pages are typically designed for visual interaction. In order to support visual interaction, Web pages are designed to consist of multiple segments with different functionalities, such as: main content, navigation bar, menu list, advertisements, etc. Recent research has shown that Web pages can be sub-divided into smaller segments. This process is known as Web page segmentation. The goal of Web page segmentation is to break a large page into smaller segments, in which contents with coherent semantics are collected [1]. Web page segmentation can be very useful for different fields, for example: Web pages can be properly displayed or repurposed for mobile devices [2–4], duplicate Web pages can be detected [5, 6], information retrieval systems can use such implicit information to provide better search results [7, 8], etc.

Recognizing the importance of Web page segmentation, numerous previous works have proposed to solve this problem. These works can be roughly divided into two types: an HTML structure-based method and a visual heuristic-based method. An HTML structure-based method often transforms HTML code into a Document Object Model (DOM) tree or HTML tag tree, and divides pages based on their pre-defined syntactic structure. However, tags such as <TABLE> and <P> are used not only for content markup but also for layout structure presentation. It is difficult to obtain the appropriate segmentation granularity [7]. Visual heuristic-based approaches rely on visual cues from browser renderings. Most of the vision-based methods focus on the location, size or font features of elements. However, most of these methods involve some set of heuristics. These heuristics typically utilize many features present on a Web page. While a heuristic approach might work well on small sets of pages, it isn't suitable for large-scale sets of pages [6].

In this chapter we propose two parameters for Web page segmentation. The two parameters are Seam Degree (SD) and Content Similarity (CS). Seam Degree describes the seam degree of two adjoining blocks. Content Similarity describes the similarity of contents in two blocks. The two parameters can utilize the vision and content feature to describe the coherent degree of Web page blocks. A Web pages block may contain many smaller sub-blocks. The averaging coherent degree of sub-blocks can determine whether a block should be divided into smaller blocks. By adjusting the threshold of the two parameters, we can obtain a fine-grained page segmentation result. These two parameters do not depend on either pre-defined HTML syntactic structure or visual heuristics. We built a page segment system using the two parameters. Through empirical analysis we show that the page segment system can divide a Web page into appropriate semantic segments.

The rest of the chapter is organized as follows: Related works are reviewed in Sect. 2. Notation and problem description are introduced in Sects. 3 and 4. The seam degree and content similarity are described in Sects. 5 and 6. A segmentation method is proposed in Sect. 7. Empirical analysis and result are reported in Sect. 8. Finally, conclusion and future work are given in Sect. 9.

2 Related Work

In the past few years, there has been plenty of work on automatic Web page segmentation. A Good survey of works on Web information extraction can be found in [9]. The page segmentation solutions roughly fall into two categories: HTML structure-based approaches and vision-based approaches.

2.1 HTML Structure-Based Approaches

HTML source code is often transformed into DOM tree or tag tree. Chakrabarti et al. [5] proposed a graph-theoretic approach to Web page segmentation. Their approach is based on formulating an appropriate optimization problem on weighted graphs, where the weights can determine whether two nodes in the DOM tree should be placed together or apart in the segmentation. However, this algorithm needs data learning and this could be an issue in the overall automation of the process. Liu et al. [10] proposed a Gomory-Hu Tree based Web page segmentation algorithm. The algorithm firstly extracts vision and structure information from a web page to construct a weighted undirected graph, whose vertices are the leaf nodes of the DOM tree and the edges represent the visible position relationship between vertices. Then it partitions the graph with a Gomory-Hu tree based clustering algorithm. Hattori et al. [11] proposed a Web page segmentation method which utilized both content-distance and page layout information. The content-distance depends on the relative HTML tag hierarchy, and layout analysis is only based on the HTML tag. However the layout information of an HTML tag does not always correspond to the actual layout of a Web page.

2.2 Vision-Based Approaches

Vision-based approaches rely on visual cues from browser renderings. Most of the vision-based methods focus on the location, size or font features of elements. Cai et al. [12] proposed a Vision-based Page Segmentation (VIPS) algorithm to divide a web page into semantic segments. They consider that each DOM node corresponds to a block. Each node is assigned a value (Degree of Coherence) to indicate how coherent the content is in the block. However, the VIPS algorithm depends on the visual cues, which are only fit for a small set of Web pages. Guo et al. [13] proposed to use visual renderings of the web page provided by Mozilla. The authors indicate that information about spatial locality is most often used to cluster, or draw boundaries around groups of items in a web page, while information about presentation style similarity is used to segment or draw boundaries between groups of items. Xiang et al.

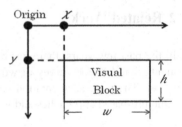

Fig. 1 The absolute coordinate and size of a visual block

[4] proposed that a web page is considered as a composition of basic visual blocks
and separators. Therefore, their algorithm focuses on first identifying the blocks and
then discovering the separators between these blocks.

Besides the two major categories, there are several other methods [1, 5, 14].
Due to paucity of space we don't introduce these methods. Our work is closed to
VIPS. However VIPS utilizes the visual cues which cannot be suitable for every
page. Moreover, these visual cues cannot correctly indicate the difference between
different semantic segments. Instead of analyzing the visual cues, we utilize seam
degree and content similarity to indicate how coherent the content is in the block.

3 Notations

A web page is made up of finite blocks. We also call these blocks visual block or
block for short. We consider a visual block as a visible rectangular region on a web
page. The definition of a visual block is as follows:

Definition 3.1 Visual block $VB = (E, R)$, where E is an Element object that is
defined by the HTML DOM based on W3C standard, and R represents the visible
rectangular region where VB is displayed in the web page.

According to W3C standard, the Element object of the DOM represents an element
in the HTML document. The details of Element object can be found in the official
W3C website.[1] The Element object not only contains the attributes of an HTML
element, such as "tagName", "id", "value" etc., but also contains the properties
defined by the DOM, such as "childNodes", "nextSibling", etc. Besides the DOM
Element, the other parameter is the visual rectangular region $R = (x, y, w, h)$ as
shown in Fig. 1. Here x is the horizontal coordinate, y is the vertical coordinates of
top-left point of visual block, w is the width, and h is the height of the visual block.
Sometimes they are written as $Rx(VB)$, $Ry(VB)$, $Rw(VB)$ and $Rh(VB)$ when only one
parameter needs to be mentioned. According to the definition of visual block, not all
HTML elements have their corresponding visual blocks. The elements that are not
visible such as <head>, <script>, <meta>, etc. and the elements whose "display"

---

[1] http://www.w3.org/standards/techs/dom#w3c_all

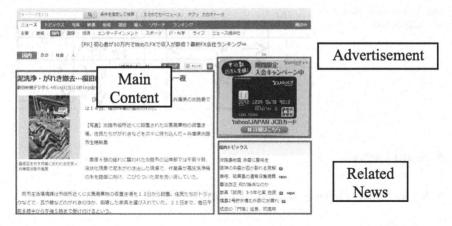

Fig. 2 An example of different semantic segment.

property is "none" or "hidden" property is "true" are not considered as a visual block in this chapter.

Definition 3.2 For two given visual blocks $VB_1 = (E_1, R_1)$ and $VB_2 = (E_2, R_2)$, if E_1 is a child node of E_2, then VB_1 is the child of VB_2.

Definition 3.3 If a visual block $VB = (E, R)$ does not have any children, then VB is a leaf visual block, denoted $VB : leaf$.

4 Problem Discussion

The purpose of our work is to break a large page into smaller segments, in which contents with coherent semantics are collected.

A Web page can be considered as a large block, which consists of several child blocks with different semantics, such as: main content, navigation bar, menu list, advertisements, etc. These segments have different functions and visual character-istics. For example, in a news site, a long text may be the main content; a link list may be the related news list; a big picture may be an advertisement, etc. Figure 2 shows the example. We can utilize the visual and content difference to indicate how coherent the child blocks are. If the coherent degree of child blocks is high, then the block should not be divided, otherwise it should be divided further. Therefore the issue of page segmentation can be seen as an issue of calculating the coherent degree of child blocks in each block. In this chapter, we introduce two parameters to describe the coherent degree, they are: the Seam Degree and Content Similarity. In the next section, we will introduce the two parameters in detail.

Fig. 3 Two examples of adjoining blocks

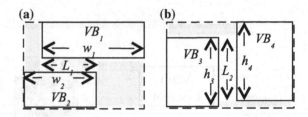

5 Seam Degree

5.1 The Seam Degree of Two Adjoining Blocks

The seam degree is used to describe how close two adjoining blocks are. First, we give the definition of adjoining blocks.

Definition 5.1 For two given visual blocks VB_1 and VB_2, let's assume that $Ry(VB_1) + Rh(VB_1) \leq Ry(VB_2)$. If the following conditions are satisfied:

(1) $Max\{Rx(VB_1), Rx(VB_2)\} \leq Min\{Rx(VB_1) + Rw(VB_1), Rx(VB_2) + Rw(VB_2)\}$;
(2) There is NOT a VB_i which is between VB_1 and VB_2.

We define VB_1 and VB_2 are adjoining in the vertical direction. Similarly, we can also define two visual blocks are adjoining in the horizontal direction (we skip over it here). If VB_1 and VB_2 are adjoining in the vertical direction or horizontal direction, we define VB_1 and VB_2 are adjoining blocks. Figure 3 shows two examples of adjoining blocks.

In Fig. 3, VB_1 and VB_2, VB_3 and VB_4 are adjoining blocks. The dotted rectangles are the minimum rectangles that cover the two blocks in Fig. 3a, b. L_1 and L_2 are the seam length of the two adjoining blocks, w_i is the width of VB_i, and h_i is the height of VB_i. Intuitively, we consider VB_3 and VB_4 are closer than VB_1 and VB_2. This is because VB_3 and VB_4 can almost fill up the minimum rectangle, but VB_1 and VB_2 cannot fill up it. The gray regions indicate the areas that are not filled up in Fig. 3. It is known that each segment has a corresponding rectangle appearing in the page. In other words, VB_3 and VB_4 are more likely to be a segment, but VB_1 and VB_2 cannot be considered as a segment. We utilize seam degree to describe the visual coherent degree. The definition of seam degree is given as follows:

Definition 5.2 For two given visual blocks VB_1 and VB_2, if VB_1 and VB_2 are adjoining in vertical direction. The seam degree $SD(VB_1, VB_2)$ can be calculated as in formula (1):

$$SD(VB_1, VB_2) = \frac{SeamLength\,(VB_1, VB_2)^2}{Rw\,(VB_1) \times Rw\,(VB_2)} \tag{1}$$

where $SeamLength\,(VB_1, VB_2)$ represents the seam length of VB_1 and VB_2, and $Rw(VB_i)$ represents the width of VB_i. Similarly, if VB_1 and VB_2 are adjoining in horizontal direction, The seam degree $SD(VB_1, VB_2)$ can be calculated as in formula (2):

$$SD(VB_1, VB_2) = \frac{SeamLength\,(VB_1, VB_2)^2}{Rh\,(VB_1) \times Rh\,(VB_2)} \qquad (2)$$

where $Rh\,(VB_i)$ represents the height of VB_i.

$SD\,(VB_1, VB_2)$ is between 0 and 1. Since the seam degree is based on the visual information of blocks, it can indicate the visual coherent degree of adjoining blocks.

5.2 The Averaging Seam Degree of Adjoining Child Blocks in a Block

If a block has child blocks, the averaging seam degree of adjoining child blocks can indicate the visual coherent degree of the content in the block. For a given visual block VB, the set of child blocks in VB is $Child\,(VB) = \{b_1, b_2, \ldots, b_n\}$. If two child blocks are adjoining, we count 1 pair. Let us assume that there are n pairs of adjoining child blocks. The averaging seam degree $AvgSD(VB)$ can be calculated as in formula (3);

$$AvgSD(VB) = \frac{\sum SD(b_i, b_j)}{n} \qquad (3)$$

where b_i and b_j are adjoining child blocks.

$AvgSD(VB)$ degree is also between 0 and 1. If it is closer to 0, the visual coherent degree of child blocks is lower. If it is closer to 1, the visual coherent degree of child blocks is higher.

6 Content Similarity

6.1 The Content Vectors of a Block

As mentioned before, segments with different semantics always have different types of contents. For example, a navigation bar has a list of short link text; an advertisement has a big picture; a user registration form has some text boxes, pull-down menus, buttons, etc. If the contents of two blocks are similar, the two blocks have a high content coherent degree. We introduce the Content Similarity to describe the content coherent degree. We roughly classify the contents into four categories:

(1) Text Contents (TC): all the text falls into this category, except the text that contains a hyper link.

(2) Link Text Contents (LTC): the text that contains a hyper link can be classified into this category.
(3) Image Contents (IMC): this category contains pictures, photos, icons, etc.
(4) Input Contents (INC): this category includes elements that can accept user input, such as: text box, radio button, pull-down menus, etc.

For a given VB, the content set is $C = \{c_1, c_2, \ldots, c_n\}$. First, the contents are classified into the four categories mentioned above. Then four types of content sets can be obtained, denoted $TC = \{tc_1, tc_2, \ldots, tc_o\}$, $LTC = \{ltc_1, ltc_2, \ldots, ltc_p\}$, $IMC = \{imc_1, imc_2, \ldots, imc_q\}$, and $INC = \{inc_1, inc_2, \ldots, inc_r\}$. Obviously, TC, LTC, IMC and INC are the subsets of C. If one of the content subsets is $\emptyset$, it means that VB does not contain the contents of the corresponding type. We use $Area(c_i)$ to represent the area of the corresponding block of c_i. If c_i is a text content or link text content, we approximately calculate the area as in formula (4):

$$Area(c_i) = Length(c_i) \times FontSize(c_i)^2 (c_i \in TC \cup LTC) \qquad (4)$$

where $Length(c_i)$ represents the length of text or link text, $FontSize(c_i)$ represents the font size of text or link text.

According to the area of contents, the four content subsets can be sorted from large to small area. By utilizing the sorted content subsets, four content area vectors can be obtained, denoted V_{tc}, V_{ltc}, V_{imc} and V_{inc}. The values of elements in the four vectors are the areas of corresponding contents. After the content vectors are determined, the content similarity of two blocks can be calculated.

6.2 The Content Similarity of Two Blocks

If the content vectors of two given blocks are determined, the similarity of each content area vector can be calculated. There are many algorithms to calculate the similarity of two vectors, of which the cosine similarity is a simple and efficient algorithm [15]. Here we take the vector of the text content as an example to explain the calculation of cosine similarity. For two given blocks VB_1 and VB_2, their text content area vectors are $V_{tc\_1} = (u_1, u_2, \ldots, u_m)$ and $V_{tc\_2} = (v_1, v_2, \ldots, v_n)$. Let us assume that $V_{tc\_1} \neq \emptyset$, $V_{tc\_1} \neq \emptyset$, and $n > m$. Because the cosine similarity requires that the two vectors must have the same number of elements, we need to add $(n - m)$ elements whose value are 0 into V_{tc_1}, denoted $V'_{tc\_1} = (u_1, u_2, \ldots, u_m, u_{m+1}, \ldots, u_n)$. The cosine similarity of V'_{tc_1} and V_{tc_2} can be calculated as in formula (5):

$$Cos(V'_{tc\_1}, V_{tc\_2}) = \frac{\sum_{i=1}^{n} u_i \times v_i}{\sqrt{\sum_{i=1}^{n} (u_i)^2} \times \sqrt{\sum_{i=1}^{n} (v_i)^2}} \qquad (5)$$

If both V'_{tc_1} and V_{tc_2} are $\emptyset$, Cos (V'_{tc_1}, V_{tc_2}) is ill-formed. In this case, we define the $Cos(V'_{tc\_1}, V_{tc\_2})$ to be zero. Similarly, the cosine similarity of other content area vectors (including V_{ltc}, V_{imc} and V_{inc}) can also be determined.

Additionally, the four types of contents have different weight in VB_1 and VB_2. Also, we take the text content as an example to explain the calculation of weight. For two given blocks VB_1 and VB_2, their text content area vectors are $V_{tc\_1} = (u_1, u_2, \ldots, u_m)$ and $V_{tc\_2} = (v_1, v_2, \ldots, v_n)$. The weight of text content can be calculated as in formula (6):

$$Weight\,(Tc) = \frac{\sum\limits_{i=1}^{m} u_i + \sum\limits_{j=1}^{n} v_j}{Area\,(VB_1) + Area\,(VB_2)} \tag{6}$$

where the $Area(VB_i)$ represents the total area of all contents in VB_i. It means that the greater area of the corresponding type of contents is, the higher its weight will be.

After the cosine similarity and weight of each content area vector are determined, the content similarity $CS(VB_1$ and $VB_2)$ of VB_1 and VB_2 can be calculated as in formula (7):

$$CS\,(VB_1, VB_2) = \sum Weight_i \times Cos_i \tag{7}$$

where $Weight_i$ represents the weight of four types of contents, and Cos_i represents the cosine similarity of four types of contents area vectors.

$CS(VB_1, VB_2)$ is between 0 and 1. Since the content similarity is based on the content information of blocks, it can indicate the content coherent degree of blocks.

6.3 The Averaging Content Similarity of Adjoining Child Blocks in a Block

Similar to the averaging seam degree, if a block has child blocks, the averaging content of adjoining child blocks can indicate the content coherent degree of the child blocks in the block. It should be noted that only the content similarity of adjoining child blocks is considered. For a given visual block VB, the set of child blocks in VB is $Child(VB) = \{b_1, b_2, \ldots, b_n\}$. If two child blocks are adjoining, we count 1 pair. Let us assume that there are n pairs of adjoining child blocks. The averaging content similarity $AvgCS(VB)$ can be calculated as in formula (8);

$$AvgCS(VB) = \frac{\sum CS(b_i, b_j)}{n} \tag{8}$$

where b_i and b_j are adjoining child blocks.

AvgCS(*VB*) is also between 0 and 1. If it is closer to 0, the content coherent degree of child blocks is lower. If it is closer to 1, the content coherent degree of child blocks is higher.

7 Page Segmentation Based on Seam Degree and Content Similarity

Based on the seam degree and content similarity, we propose a page segmentation method. In order to divide a page into segments, a page needs to be transformed into a DOM tree. The nodes that will not appear in the Web page should be pruned, such as the nodes whose tags are <SCRIPT>, <META>, <STYLE>, etc, and the nodes whose height or width is zero. Also, we need to get the visual information of each node by utilizing the APIs of browsers. This is because the DOM nodes do not contain the absolute coordinate. In this way, the corresponding block of each DOM node can be determined.

For a given block, the averaging seam degree and content similarity of its adjoining child blocks are calculated. We introduce two thresholds α and β to determine whether the node should be divided or not. The segmentation algorithms are as follows:

Step 1: For a given block, if the averaging seam degree and content similarity of its adjoining child blocks is less than α, then the block should be divided.

Step 2: For a given block, if the averaging seam degree and content similarity of its adjoining child blocks is greater than α, and the averaging content similarity of its adjoining child blocks is less than β, then the block should be divided.

Step 3: For a given block, if it does not satisfy the Step 1 and Step 2, then the block should not be divided.

If a given block that does not contain any child block, we define both its averaging seam degree and content similarity are one. If a given block that contains only one child block, we define both its averaging seam degree and content similarity is zero. Our method is a top-down algorithm. We calculate the averaging seam degree and content similarity from the root block. If a block should be divided, its child blocks will be checked further. If a block should not be divided, it will be pushed into a segment array and its child blocks will not be checked any more. Finally, all the segments can be determined.

8 Experiment and Analysis

The goal of the experiment is to determining the optimal thresholds of Seam Degree (AvgSD) and Content Similarity (AvgCS) and comparing the proposed method with VIPS. In order to achieve this goal, we collected the web pages from internet. We randomly selected 10 keywords from 10 different categories of Yahoo Chiebukuro.

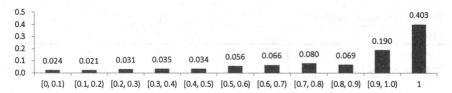

Fig. 4 Distribution of AvgSD(b)

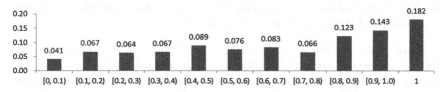

Fig. 5 Distribution of AvgCS(b)

We submitted 10 queries to Google, from which we randomly collected 10 pages from the search results as test pages. As a result, we collected 100 web pages which are from 88 different web sites.

Firstly, we determine the optimal thresholds of AvgSD and AvgCS. Within the 100 web pages, there are 32842 blocks. For each block b, we calculate the AvgSD and AvgCS of each block denoted AvgSD(b) and AvgCS(b). Since both AvgSD(b) and AvgCS(b) are within interval [0, 1], we divided the interval [0, 1] into 11 smaller interval, they are $\{[0, 0.1), [0.1, 0.2)...[0.9, 1), [1, 1]\}$. It should be noted that [1, 1] is a single interval. After the intervals are determined, we can calculate the distribution of AvgSD(b) and AvgCS(b). Figures 4 and 5 shows the distribution of of AvgSD(b) and AvgCS(b). In the two figures, the horizontal axis is the intervals and the vertical axis is the percentage of blocks in each interval.

In Fig. 4, it is obvious that the distribution of AvgSD(b) with [0, 0.9) are nearly the same. However, within interval [0.9, 1] the distribution of AvgSD(b) increased steeply. Moreover, Fig. 4 shows the blocks whose AvgSD are greater than 0.9 makes up 59.3 % of the total. Therefore AvgSD(b) = 0.9 can be considered as a breakpoint and we set $\alpha = 0.9$ as the optimal threshold of Seam Degree.

Similarly, in Fig. 5, it is obvious that the distribution of AvgCS(b) with [0, 0.8) changes little. However, within interval [0.8, 1] the distribution of AvgCS(b) increased steeply. Moreover, Fig. 5 shows the blocks whose AvgCS are greater than 0.8 makes up 45% of the total. Therefore AvgCS(b) = 0.8 can be considered as a breakpoint and we set $\beta= 0.8$ as the optimal threshold of Content Similarity.

After the optimal thresholds are determined, we compared the proposed method with VIPS [12]. There are three reasons why we chose VIPS as the baseline of our experiment:

(1) No quantitative analysis to indicate the best page segmentation method. Most analyses are based on subjective evaluation or specified data set.
(2) VIPS is the most common baseline for evaluating page segmentation method.

Table 1 Comparing with VIPS

Level	1	2	3	4	5
VIPS (%)	9	16	19	25	31
Proposed method (%)	1	2	9	18	70

(3) VIPS is a vision-heuristic-based method, and it can determine whether the proposed formulated visual features are effective or not.

Based on the experiment result, we lets α be 0.9 and β be 0.8, and segment the 100 pages using our method. We conducted a questionnaire experiment to compare the proposed method and VIPS. There are 14 participants who evaluated the segment results. We use 1~5 five levels to evaluate the results where 5 is best and 1 is worst. Table 1 shows the questionnaire results. The proposed method is perfectly suitable for 70% web pages. It means that the formulated visual features are more universally effective than vision-heuristics. If both level 4 and level 5 results are regarded as acceptable results, the proposed method can be suitable for 88% web page. We also analyzed the pages whose segmentation result are level 1 and level 2. The level 1 and level 2 results were due to "Text Only" web pages. In this case, the content similarity would be ineffective. However, this kind of web pages is only 3%. Therefore, we can conclusion that the proposed method is more effective than VIPS.

9 Conclusion and Future Work

In this chapter, we proposed two parameters seam degree and content similarity to indicate the coherent degree of a page block. The seam degree is based on the visual information of blocks, therefore it can indicate the visual coherent degree of adjoining blocks. The content similarity is based on the content information of blocks, therefore it can indicate the content coherent degree of blocks. Instead of analyzing pre-defined heuristics or visual cues, our method utilized the visual and content coherent degree to determine whether a page block should be divided into smaller blocks. We also proposed a page segmentation method using these two parameters. An experiment was conducted to determine the relationship between the two parameters and the number of segment result. The empirical results also show that our segmentation method is effective to segment a page into different semantic parts.

However our method cannot identify recurrent blocks. For example, in the search result page of Amazon, each product record has an independent semantic. Since they have similar contents, they are probably not divided into different segments. In the future, we are planning to solve this problem and improve the segmentation results.

References

1. P. Xiang, X. Yang, Y. Shi, Web page segmentation based on gestalt theory, in *Multimedia and Expo 2007 IEEE International Conference (ICME)*, pp. 2253–2256 (2007)
2. X. Yin, W.S. Lee, Using link analysis to improve layout on mobile devices, in *Proceedings of the Thirteenth International World Wide Web Conference*, pp. 338–344 (2004)
3. X. Xie, G. Miao, R. Song, J. Wen, W. Ma, Efficient browsing of web search results on mobile devices based on block importance model, in *Proceedings of the Third IEEE International Conference on Pervasive Computing and Communications*, pp. 17–26 (2005)
4. P. Xiang, Y. Shi, Recovering semantic relations from web pages based on visual cues, in *Proceedings of the 11th international conference on Intelligent user interfaces(IUI'06)*, pp. 342–344 (2006)
5. C. Kohlschutter, W. Nejdl, A densitometric approach to web page segmentation, in *Proceeding of the 17th ACM conference on Information and knowledge management, CIKM '08*, pp. 1173–1182 (2008)
6. D. Chakrabarti, R. Kumar, K. Punera, A graph-theoretic approach to webpage segmentation, in *WWW'08: proceeding of the 17th international conference on World Wide Web*, pp. 377–386 (2008)
7. A. Madaan, W. Chu, S. Bhalla, VisHue: web page segmentation for an improved query interface for medlineplus medical encyclopedia. Databases Networked Inf. Syst. **7108**, 89–108 (2011)
8. K.S. Kuppusamy, G. Aghila, Multidimensional web page segment evaluation model. J. Comput. **3**(3), 24–27 (2011)
9. Y. Yesilada, Web page segmentation: a review. Technical Report. University of Manchester and middle east technical university northern cyprus campus (2011) (Unpublished)
10. X. Liu, H. Lin, Y. Tian, Segmenting webpage with Gomory-Hu tree based clustering. J. Softw. **6**(12) (2011)
11. G. Hattori, K. Hoashi, K. Matsumoto, F. Sugaya, Robust web page segmentation for mobile terminal using content-distances and page layout information, in *Proceedings of the 16th international conference on World Wide Web (WWW '07)*, pp. 361–370 (2007)
12. D. Cai, S. Yu, J. Wen, W.g Ma, VIPS: a vision based pagesegmentation algorithm. Technical Report MSR-TR-2003-79, Microsoft Research (2003)
13. H. Guo, J. Mahmud, Y. Borodin, A. Stent, I.V. Ramakrishnan, A general approach for partitioning web page content based on geometric and style information, in *Proceedings of the International Conference on Document Analysis and Recognition*, pp. 929–933 (2007)
14. H. Sano, R.M.E. Swezey, S. Shiramatsu, T. Ozono, T. Shintani, A web page segmentation method by using headlines to web contents as separators and its evaluations. Int. J. Comput. Sci. Netw. Secur. **13**(1), 1–6 (2013)
15. S. Amit, Modern information retrieval: a brief overview. Bull. IEEE Comput. Soc. Tech. Committee Data Eng. **24**(4), 35–43 (2001)

Improving Particle Swarm Optimization Algorithm and Its Application to Physical Travelling Salesman Problems with a Dynamic Search Space

Benoît Vallade and Tomoharu Nakashima

Abstract This chapter addresses an improvement idea for the Particle Swarm Optimization Algorithm (PSO) and its implementation on a Traveller Salesman Problem based competition. As a search algorithm, the PSO is used to tune a set of parameters, which usually take their values in static search spaces. This chapter proposes a solution to use effectively the PSO algorithm on optimization problems using parameters which take their values in dynamic space.

Keywords Algorithm of non-deterministic search · Particle swarm optimization algorithm · Dynamic search space · Path finding algorithm · Traveller salesman problem.

1 Introduction

Nowadays, One of Artificial Intelligence major problems is the optimization problem. Such problems consist of finding the best combination of the parameter values. A common way to solve such problems is to use the algorithms of non-deterministic search. There are various kinds of search algorithm [1] such as tabu algorithm, genetic algorithm, particle swarm optimization (PSO) algorithm and others.

Among all these algorithms, this chapter focuses on the particle swarm optimization algorithm also called the PSO algorithm. The concept of the PSO algorithm is based on the simulation of a simplified social model and more particularly on the animals flocking [2]. Its conception follows some standard which have evolved

B. Vallade (✉) · T. Nakashima
Department of Computer Science and Intelligent Systems, Osaka Prefecture University, Osaka, Japan
e-mail: valladeben@cs.osakafu-u.ac.jp

T. Nakashima
e-mail: tomoharu.nakashima@kis.osakafu-u.ac.jp

R. Y. Lee (ed.), *Applied Computing and Information Technology*,
Studies in Computational Intelligence 553, DOI: 10.1007/978-3-319-05717-0_8,
© Springer International Publishing Switzerland 2014

Table 1 Set of variables' static search spaces

	Minimum	Maximum
A	−5	5
B	2	6
C	−10	−5
D	0	10
E	−2	5

overtime [3]. Similarly to the other algorithms of non-deterministic search, the standard PSO algorithms allow to tune a set of parameters, which take their values in a static search space. This means that any time during the optimization, the search space of each variable remains the same. However, some optimization problems as the robot motion optimization use a set of variables, which take their values in dynamic search spaces [4]. This means that the search spaces of the variables may vary during the optimization.

This chapter presents principally our solution to improve the efficiency of the PSO algorithm on these problems that involve variables with a dynamic search space. First, in the next section, we will describe the global concept of the standard versions of the PSO algorithm. Next, Sect. 3 explains in details the particularities of these dynamic problems and the algorithm's improvement used to solve them. Section 4 gives the results of some experiments which compare the efficiency of both algorithms, standard and new, on these problems. Then, we will analyze the utility of the proposed PSO algorithm on real world problems. For this we begin by determining for which kind of problems this algorithm can be well designed. In addition, we will use it on a competitive problem of a physical traveler which correspond to this kind problem and evaluate its efficiency against other kinds of solutions to prove its effectiveness. Finally, we conclude this chapter by summarizing the proposed method, the performance the new algorithm, and its usability on real world problems.

2 Standard Particle Swarm Optimization

2.1 Concept Overview

As introduced before, the PSO algorithm is an non-deterministic search algorithm. This means that it searches for the best combination of values for a set of variables. As the Table 1 shows, the variables take their values in search spaces defined by a minimal and a maximal values. These limits are given by the user and will take constant values.

Consequently, another way to represent an optimization problem is to consider a solution as a position in a search space defined by crossing all the individual parameters' search space.

The special feature of the PSO algorithm is that its concept is based on the birds flocking [2]. During their feeding time, multiple birds evolved in the same space and

Fig. 1 Birds' movements
according to best positions
knowledge

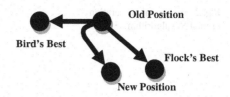

search for the position where there is the biggest quantity of foods. In the course
of their search, each bird always remembers the position where they have found the
biggest quantity of food. In addition, as the birds follow a social behaviour inside
the flock, they also share the best position found by the whole flock. Finally, as
shown in Fig. 1, each bird adapts its movements in the search space according to
these knowledge.

This social behaviour is used by the PSO algorithm as a conceptual idea to generate
new solutions and optimize the set of parameters. In this transposition, the birds will
be called particles, the flock will be the swarm and the quantity of foods of a position
will correspond to the quality of the solution.

2.2 Standard Algorithm

Our research is based on 2011s version of the standard PSO algorithm described in the
chapter of Maurice Clerc [3], with the particularity of not using the neighbourhood
system (in case of neighbourhood system, the particles are grouped in teams and
they share the information about the best position found by all the team's member
only inside the team, in our case there is only one big team, which correspond to
the whole swarm). This part of the chapter gives some details about this version of
standard PSO algorithm.

2.2.1 Particle's Components and Algorithm

As explained in the previous part on the birds flocking transposition, a swarm of
particles is included in the search space. Each particle is aware of:

- Its Position (initialized randomly in the search space).
- Its Velocity (initialized randomly in the search space).
- Its Best Position ever found (initialized as the first particle's position).
- The Swarm's Best Position (initialized by comparing all the quality Particle's first
  position).

It should be noted that in the 2011 version, the initialization of the positions' and
velocities' values are randomly generated, parameter by parameter.

Each iteration of the optimization, these particles' attributes are updated following
the process described in Fig. 2:

Fig. 2 Process of an iteration
of the PSO algorithm

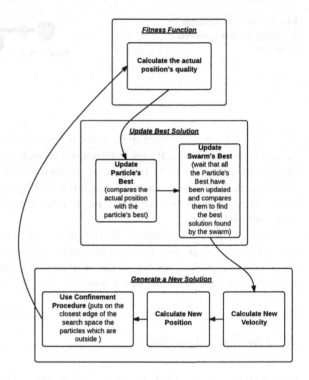

2.2.2 Evolution Rules

Below are given the equations used for the velocity and position update and all the other details useful to the implementation of the 2011 version of the PSO algorithm.

Swarm size and initialization:

In the 2011 version the swarm size (number of particles) will be user defined, with 40 as suggested value. The initialization of the particles' attributes will be conducted as described before.

Calculate Velocity and Position:

First the velocity will be updated by using the following equations:

$$G_i = x_i + c \times (\frac{p_i + l_i - 2 \times x_i}{3}) \tag{1}$$

$$H_i(G_i, \|G_i - x_i\|) \tag{2}$$

$$v_i(t+1) = w \times v_i(t) + x_i'(t) - x_i(t) \tag{3}$$

In these equations:

- $c = 1.193$ and $w=0.721$ (constants).
- x_i (or $x_i(t)$) is the actual position value of the ith particle.
- p_i is the particle's best position of the ith particle.
- l_i is the swarm's best position of the ith particle.
- G_i is the centre of gravity of the three points x_i, p_i and l_i.
- H_i is the hyper sphere of centre G_i and radius $G_i - x_i$.
- $x_i'(t)$ is a position randomly choose in the hyper sphere H_i.
- $v_i(t)$ is the actual velocity of the ith particle.
- $v_i(t+1)$ is the new velocity of the ith particle.

Next, the position is updated by using the equation:

$$x_i(t+1) = w \times v_i(t) + x_i'(t) \tag{4}$$

In cases where $l_i = p_i$, the following gravity centre equation is used for the velocity updating:

$$G_i = x_i + c \times (\frac{p_i - x_i}{2}) \tag{5}$$

Confinement:

But, sometimes, the new position of the particle is out of the search space. In those cases, the algorithm uses a confinement procedure, which moves the particle on the closest edge of the search space. This movement is conducted by replacing the value of each parameter of the position by the closest corresponding parameter's search space limits, min or max. Finally, the velocity forced to the following value:

$$v_i(t+1) = -0.5 \times v_i(t+1) \tag{6}$$

Particle's and Swarm's Best:

To finish the quality of the new position is calculated by using the fitness function. As said before, this function depends on the problem and is defined by the user. Its results will be used to compare the different positions found by the algorithm. During a first comparison the value of the particle's best position is updated in function of the previous one and of the actual position. In order to do the second comparison, the algorithm waits that all the particles' best of the swarm are updated. All the particles' best position of the swarm will be compared to determine which the swarm's best position is, and this knowledge will be shared with all the swarm's particles.

3 PSO in Dynamic Search Spaces

This section discusses the topic of the problems based on set of parameters with dynamic search spaces. And then the proposed solutions to deal with such problems and through the changes made on the standard PSO algorithm are explained.

Table 2 Set of variables' dynamic search spaces

	Minimum	Maximum
A	0	5
B	$-A$	A
C	$-5*A+B$	$+5*A+B$
D	-15	$2*A$
E	-20	10

Fig. 3 Dynamic search space representation in a two-parameter optimization problem

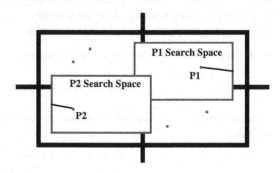

Table 3 Set of variables' maximal search spaces

	Minimum	Maximum
A	0	5
B	-5	5
C	-30	30
D	-15	10
E	-20	10

3.1 Dynamic Search Space Problems

Contrarily to the previous standard types of problems, which used parameters taking their values in static search spaces, some problems are based on dynamic search spaces. In these kinds of problems, the search spaces limit of some parameters depends on the value of other parameters. Table 2 gives an example of such a set of parameters:

In those kinds of problems, the search spaces limits depends on the value of the parameters and consequently on the position of the particle. Finally, as shown in Fig. 3, all the particles evolved in a different search space, which changes in function of the particle's position. However, a maximal space search can be created by using the maximum value possible for each parameter's maximum limit and the minimum value possible for the parameter's minimum limit.

Table 3 uses the Table 2's example to create the corresponding maximal search space.

This space consequently contains all the individual search spaces possible. By individual search space, we mean the search spaces which are created by using the position value. But it also contains positions which are not included in these individual search spaces. To keep the precedent example (Table 2 and 3), the position of coordinate $(0; -5; 0; 0; 2)$ is available in the maximal search space but not in individual spaces. As the search spaces limits are user-defined, we will call these positions, "uninteresting positions".

3.2 Dynamic Search Space PSO Concept

To solve such problems, there are two options. The first one is to apply the optimization on the maximal search space. As described above, this space includes all the search spaces possible and has the particularity to be static. The advantage of such a solution is that, as the search space is static, the standard PSO algorithm described before can be used. The disadvantage is that the optimization will also be conducted on uninteresting positions. This may result in loss of time and a final optimization position not intended by the user.

The second solution is to use individual search spaces. This means that each particle will have its own search space and this one will change as the same time as the particle move. This solution avoids the search on the uninteresting positions but implies some modifications to the standard algorithm.

To be able to improve the efficiency of the PSO algorithm in case of dynamic search space problems, we chose the second solution. Contrary of the standard algorithm, our improved algorithm will use particles which have their own search spaces. The following part explains the necessary changes on the standard algorithm.

3.3 Dynamic Search Space PSO Modifications

This part discusses about the problem faced by the standard algorithm resulting from our choice and about the possible modifications to avoid it.

3.3.1 Problem

During the initialization step as well than during the confinement methods, the new value of the position will be generated parameter by parameter. A random value will be generated between the parameter's search space limits for the initialization and the closest limit will be searched for the confinement. But, by using dynamic search spaces, these limits values will depend on others parameters' values. Also, the algorithm would not be able to generate a parameter's value if its limits have not ever been defined. That is why the parameters' values need to be defined in the good order.

3.3.2 Modification

This order will of course be based on the links between the parameters. The parameter A is linked with the parameter B if value of B is required to calculate the limits of A's search space.

In the aim to represent these links, we chose to use an acyclic graph representation; they have the particularity to allow multiple roots and multiple parents for a same child. Of course we can't allow cycle due to the impossibility to generate a position value if the parameters are linked through a cycle. In this case, the first proposed methods using the maximum search space should be used.

Our implementation uses a unique root, which does not correspond to any parameter, but it allows us to insert all the parameters in the same graph.

Each parameter is represented by a node which is linked to other nodes following the parameters links. The node A is parent of the node B if the parameter A needs the value of B. The nodes will be sorted in depth layers, a node will also be in the layer under the layer of its deepest parents. The direct leafs are inserted in the deepest layer of the graph.

Finally, the algorithm operates on the same way than the standard one with the following modification. During the initialization step to generate the position value and during the confinement procedure to correct the position value, the algorithm generates and corrects the value following the order decided by using the graph. This path corresponds to a back breadth-first search of the graph. This means that we start from leafs and we head to the root of the graph by visiting each node of a layer before to go to the upper one. Node includes calculate its search space limits and generate its value (or search the closest limits for the confinement method). Figure 4 is the acyclic graph representation of the example given in Table 2. It also shows the calculation path (dotted arrows):

4 Experiments

In order to compare the efficiency between the standard and the new algorithm on dynamic search spaces problems, we set up two experiments. The problem is that there are no such problems in literature [5]. Consequently, we cannot compare it with the other algorithms' results but have to create our own artificial problems. This is the why the two next experiments do not correspond to any known problems and have no real correspondence with the real life. These problems have only been designed to proof the functionality and efficiency of the new algorithm compared with the standard algorithm.

4.1 Experiment 1

For this one, we will use a swarm of 40 particles and launch optimization of 500 iterations. We created three optimization problems very simple (so they will be solved

Fig. 4 Acyclic graph representation of parameters link and calculation path

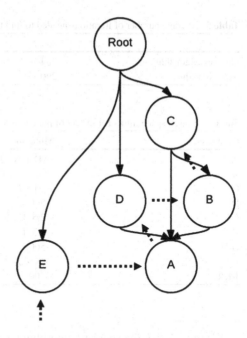

Table 4 Search space limits for the Y parameter

	New algorithm		Standard algorithm	
	Minimum	Maximum	Minimum	Maximum
A	$-(0.05*x)+0.25$	$(0.05*x)-0.25$	-49.75	49.75
B	$-(0.25*x)+1.25$	$(0.25*x)-1.25$	-248.75	248.75
C	$-(0.5*x)+2.5$	$(0.5*x)-2.5$	-497.5	497.5

in 500 iterations) and compared how many iterations are required to find the optimal position.

These problems have the following characteristics:

- two dimensional problems (parameters: X , Y).
- The Y's search space limits depends on X's value.
- The optimal position is the point of coordinate (5.5; 0.01). and the fitness function calculate the distance between the particle position and the optimal position.
- X's value vary in [0 ; 1000].
- Y's vary in the limits defined by Table 4, the standard algorithm will use the maximum search space:

We repeated the optimization 10 times and took the average number of iterations needed to find the optimal point which is find when the fitness value is zero (as we calculate the distance). Table 5 regroups the results of this experiment:

Table 5 Average number of iterations needed to find the optimal position

	A	B	C
Standard algorithm	284	295	304
New algorithm	290	265	268

Table 6 Search space limits for the set of parameters

Parameters	Mininum	Maximum
A	$-(D+E+F)$	$D+E+F$
B	$-(E+F+G)$	$E+F+G$
C	$-(H+I)$	$H+I$
D	-500	500
E	-2.5^*J	2.5^*J
F	-0.5^*J	0.5^*J
G/H	-500	500
I	$-(K^*0.5)$	$K^*0.5$
J/K/L	-500	500

We remark that the standard algorithm has better results for the problem A, but becomes less efficient on B and C. This means that the new algorithm would be more efficient on big search spaces. We can explain this by the fact that as the new algorithm search space corresponds to affine subspace, bigger the space is for the standard algorithm and more there are uninteresting positions which are skipped by the new algorithm.

4.2 Experiment 2

As the previous results seems indicate that the new algorithm is more efficient on big search space we set up a second experiment to confirm. To do so we still used the same configuration for the PSO algorithm (40 particles, 500 iterations and we created a more complicated problem evolving on a bigger maximal search space. The search spaces limits of this problem are described in Table 6 and the parameters links can be visualized in Fig. 5.

It should be noted that the maximal search space is not given, but it can be easily calculated, as shown in the previous examples.

Also to be noted that the optimal position is a static position (0, 0..., 0), centre of the search space. The fitness function calculates the distance to this point, so the best quality value possible is 0.

In this experiment, the problem is too big to be solved in 500 iterations; so, we compare the quality of the solutions. The standard algorithm gave an average of 0.0027, while the new algorithm gave an average of $7.90 * 10^{-25}$. The results of

Fig. 5 Acyclic graph representation of parameters link

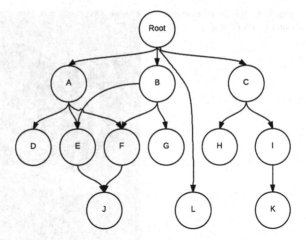

this experiment show clearly the efficiency of the new algorithm. This success is
due to the fact that the new algorithm doesn't search in the space where there is no
possibility to find a solution.

5 Physical Traveling Salesman Problem

This part discusses about the results given by the proposed PSO algorithm on a
real world problem, and more precisely on the physical travelling salesman problem
(PTSP). We compare the results of the proposed PSO with the other solutions given
by the competition organization. The PTSP is a derivation of the very known Traveler
Salesman Problem.

5.1 PTSP Competition

In the Physical Traveling Salesman Problem competition proposed by the University
of Essex [6] a "ship" will be positioned at the start position of a map and will have
to visit all the way points once in the shortest possible time. In addition there is a
countdown timer which state the time left to visit another way point.

The most important difference with the TSP is due to the physical particularity
of the problem. It implies that in addition to find the shortest way we have to take
in account the acceleration, deceleration and inertia effects to follow the calculated
itinerary.

Fig. 6 Graphical representation of a PSTP map

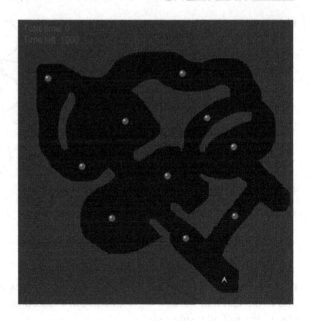

Figure 6 is the graphical representation of a map given by the competition committee. The triangle is the ship on its start point, the red dots are the way points and turn blue after have been visited. The obstacles are in grey.

5.2 Solution

To be able to experiment the proposed PSO algorithm on the PTSP competition we created a program which is divided in three modules.

The first read the map and search the shortest itinerary between all the couple of points (two way points or one way point and the starting point). It will take in account the various obstacles and record each itinerary as well than their distance.

The second module, the PSO algorithm, searches the best order to visit the way points. Its fitness function will use the distances calculated by the previous module to evaluate the global distance a solution. A solution corresponds to a sequence of way points to visit. The first parameter's value corresponds to the first way point to visit, and so on. Finally, according to our wish to visit each way point only once, we use the new algorithm rules system to force each parameter to take its value in the list of the way points which still not used. To obtain such result, we link the parameters so that the parameter number n is dependant from all the parameter from 1 to n−1. The Fig. 7 shows the acyclic graph representing the parameters linkage for a problem with 4 way points:

Fig. 7 Graphical representation of the parameter link

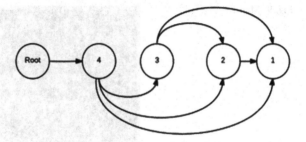

Fig. 8 Ship representation

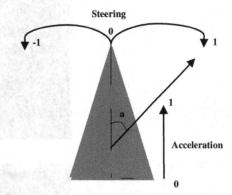

Table 7 List of ship movement instructions

Action ID	Acceleration	Steering
0	No (0)	No (0)
1	No (0)	Left (-1)
2	No (0)	Right (1)
3	Yes (1)	No (0)
4	Yes (1)	Left (-1)
5	Yes (1)	Right (1)

Finally the third module reads the itinerary file generated by the previous module and gives movement instructions to the ship. The previous figure and table (see Fig. 8 and Table 7) extract from the competition documentation [6] lists the available instructions:

5.3 Experiment

In this last part, some experiments results are given. The experiments are conduct in two different maps (see Figs. 9 and 10), one with and one without obstacles. For

Fig. 9 Map 1

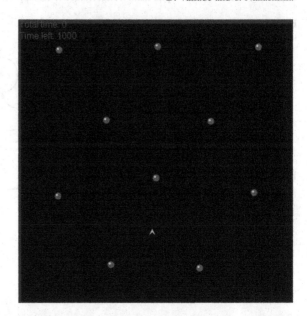

Fig. 10 Map 2

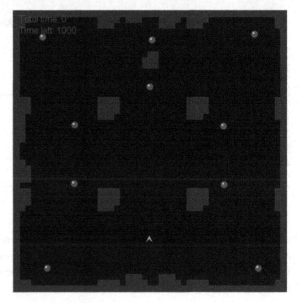

each map we launch a simulation for one of the three AI given by the competition and one for ours.

The results given by the various AI for these maps are given in the Table 8:

We can see on these two examples that our algorithm took the second place in the competition against given AI. However, there is a big difference between the first

Table 8 Experiments' results

Rank	1st	2nd	3rd	4th
Map1	MacroRandom	PSO	LineofSight	Greedy
	(1462)	(3925)	(4312)	(4675)
Map2	MacroRandom	PSO	Greedy	LineofSight
	(1383)	(4035)	(4810)	(5192)

and the second place. This difference is due to the issues encounter in managing the inertia of the ship. The ship goes out of its trajectory and loses time to come back on it. We can consequently say that these results are heavily depending on the physical aspect of the competition. However, it really proves the capacity of the proposed PSO algorithm to be use in real world problem.

6 Conclusion and Future Work

To conclude, the PSO is a very simple and easily adaptable algorithm. The standard version of space by venturing a loss of time and falling on an uninteresting result. This chapter described an efficient solution to improve its performance in this case. The experiments shown it is functional and get better results than the standard version of the PSO. It will allow to solve various real world problems by using the PSO algorithm with a minimum of modification from its standard implementation. However, there is actually no efficient solution for dynamic search space problems, where parameters are cycled linked. Our future objectives are to be able to deal with cycled graph.

References

1. H. Youssef, S.M. Sait, H. Adiche, Evolutionary algorithms, simulated annealing and tabu search: a comparative study. Eng. Appl. Artif. Intell. **141**, 167–181 (2001)
2. J. Kennedy, R. Eberhart, Particle swarm optimization. Proc. IEEE Int. Conf. Neural Networks **4**, 1942–1948 (1995)
3. M. Clerc, Standard Particle Swarm Optimisation, Technical report (2012)
4. T. Uchitane, T. Hatanaka, Applying evolution strategies for biped locomotion learning in robocup 3D soccer simulation. *Proceedings of 2011 IEEE Congress on Evolutionary Computation*, pp. 179–185 (2011)
5. M. Molga, C. Smutnicki, Test functions for optimization needs (2005). http://www.zsd.ict.pwr. wroc.pl/files/docs/functions.pdf
6. D. Perez, S. Lucas, University of Essex, Rules of the CIG PTSP 2013 Competition (2013). http:// www.ptsp-game.net/goal.php?c=league

Experimental Implementation of a M2M System Controlled by a Wiki Network

Takashi Yamanoue, Kentaro Oda and Koichi Shimozono

Abstract Experimental implementation of a M2M system, which is controlled by a wiki network, is discussed. This M2M system consists of mobile terminals at remote places and wiki servers on the Internet. A mobile terminal of the system consists of an Android terminal and it may have an Arduino board with sensors and actuators. The mobile terminal can read data from not only the sensors in the Arduino board but also wiki pages of the wiki servers. The mobile terminal can control the actuators of the Arduino board or can write sensor data to a wiki page. The mobile terminal performs such reading writing and controlling by reading and executing commands on a wiki page, and by reading and running a program on the wiki page, periodically. In order to run the program, the mobile terminal equipped with a data processor. After placing mobile terminals at remote places, the group of users of this system can control the M2M system by writing and updating such commands and programs of the wiki network without going to the places of the mobile terminals. This system realizes an open communication forum for not only people but also for machines.

Keywords M2M · Sensor network · Social network · Wiki · Java · API

T. Yamanoue (✉) · K. Oda · K. Shimozono
Computing and Communications Center, Kagoshima University, 1-21-35 Korimoto,
Kagoshima 890-0065, Japan
e-mail: yamanoue@cc.kagoshima-u.ac.jp

K. Oda
e-mail: odaken@cc.kagoshima-u.ac.jp

K. Shimozono
e-mail: simozono@cc.kagoshima-u.ac.jp

R. Y. Lee (ed.), *Applied Computing and Information Technology*,
Studies in Computational Intelligence 553, DOI: 10.1007/978-3-319-05717-0_9,
© Springer International Publishing Switzerland 2014

1 Introduction

A wiki [1] is a web site that allows the easy creation and editing of any number of interlinked web pages via a web browser and can be used as a means of effective collaboration and information sharing. Wikipedia [2] is a well-known wiki site.

If a wiki is friendly to people, it also must be friendly to machines. If a machine can read and write data on a wiki page automatically, people can obtain much more beneficial information. People also can easily control machines through the wiki page. Not only machine-to-people or people-to-machine, but also machine-to-machine communication, must be achieved easily. If such communication can be achieved, the wiki can be much more useful. For example, if a well-known wiki can be used to connect sensors in a sensor network, building one's own sensor network becomes easier.

To confirm the above presumption regarding the usefulness of wikis, we are developing a machine-to-machine (M2M) system using wiki software [3]. This system consists of mobile terminals and web sites with wiki software. A mobile terminal of the system consists of an Android [4] terminal and it may have an Arduino [5] board with sensors and actuators. We have made a mobile terminal that reads data from the Arduino board sensors and sends the sensor data to a wiki page of PukiWiki [6]. The mobile terminal also reads commands on the wiki page and controls the actuators of the Arduino board.

PukiWiki [6] software is commonly used in Japan. The authors of this chapter constructed an API for applets to allow easy and unified data input and output at a remote host. Moreover, we combined the API and the PukiWiki system by introducing a wiki tag for starting Java applets. The proposed API, which can be used to make the wiki more flexible and extensible, is referred to as the *PukiWiki-Java Connector* (PJC) [7].

The PJC enables a number of Java programs to be easily embedded in PukiWiki. We have embedded programs such as a simple text editor, a simple music editor, a simple drawing program, a programming environment, and a voice recorder in PukiWiki. One to three days was required for embedding [7]. We also have succeeded to implement a bot capturing system for network security using the PJC [8].

We also made an API to read and write data on a page of PukiWiki from an Android terminal by customizing the PukiWiki-Java connector. We call this API the *PukiWiki-Java Connector Service for Android* (PJC-S). This API enables an application of Android to read and write a page of PukiWiki easily and automatically.

We have added a data processor to the mobile terminal. This processor reads data from the sensors or wiki pages. The processor processes the data after that. And the processor outputs the processed data on a wiki page or controls actuators by the processed data. The program, which is written on a wiki page, controls this processor. The programming language of this processor is a kind of BASIC programming language. The processor has the ability of reading a kind of CSV (Comma Separated Values) format data into a table and the ability to manipulate the table like a macro program of a spreadsheet.

From the user's point of view, the program on a wiki page is directly executed when the mobile terminal read the page, just like a Java Script is executed at a web browser when the browser opened the page. The text editor of the wiki software on a web browser can edit the program of this processor. This feature allows that the user of the M2M system can exchange the program of the sensor/actuator side without going to the physical place of the sensor/actuator if the user can access the Internet and the user can use a web browser. Usually, a program in a web page, such like a Java Script or a Java Applet, can read data from or write data to the server of the web page only, whereas this processor can read data from any web servers and can write data to any server of the wiki software if the server allows it.

A mobile terminal can be placed anywhere, that is, at an Internet-accessible place as well as a place where a mobile phone can be used, if the terminal uses an Android terminal for communication between the terminal and the wiki page. Using Arduino enables many users to make their own sensors and actuators easily because Arduino is open source hardware and its programming environment is easy to use.

Many projects are carried out by groups of people. Such groups use Wiki software as a tool for collaboration. It can be used for the entire life cycle phases of the projects such as the planning, building, operating, supporting, and improving phases. Many projects for M2M systems are also carried out by groups of people so Wiki software can be used for them. Wiki software is mainly used for sharing ideas among a group of a project, at each phase of the life cycle of a project. Wiki software of the M2M system of this chapter can be used for not only sharing ideas but also prototyping, building, and improving. An idea of the group can be reflected in the M2M system immediately when the sequence of commands or the program of the wiki page is written or changed by the group. This feature can speed up these phases of the project. It is easy to put back them to their previous status when the idea was not so good because Wiki software provides this feature.

The remainder of this chapter is organized as follows. Section 2 gives an overview of the M2M system. Sections 3 and 4 illustrates details of a mobile terminal of the system and the data processor respectively. Section 5 shows an example usage of the system. Section 6 compares related work, and, finally, in Sect. 7, concluding remarks as well as future work are discussed.

2 Overview of the M2M System

Figure 1 shows an overview of the system. The system consists of mobile terminals and PukiWiki sites. A mobile terminal of the system consists of an Android [4] terminal and it may have an Arduino [5] board with sensors and actuators. A mobile terminal reads data from the Arduino board sensors and sends the sensor data to a wiki page of PukiWiki directed by commands on the wiki page. A mobile terminal of the M2M system also includes the data processor. This processor reads data from the sensors or wiki pages. The processor processes the data after that. And the processor outputs the processed data on a wiki page or controls actuators. The program that

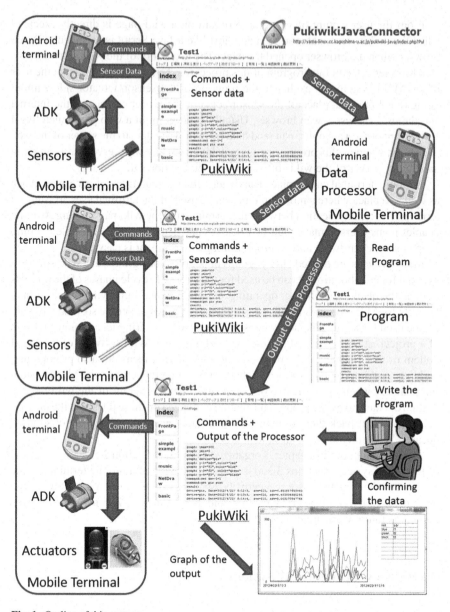

Fig. 1 Outline of this system

is written on a wiki page controls this processor. The data on the wiki page, which is acquired by a mobile terminal, can be processed by a program at the PukiWiki site and can be transformed into another data format such as a graph. This system uses a simple API for reading and writing data on a wiki page of PukiWiki software, PukiwikiJavaConnector [7], for communication between mobile terminals and wiki sites.

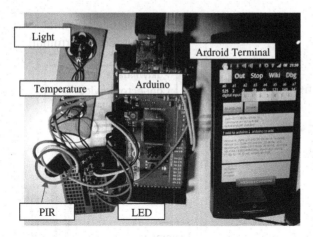

Fig. 2 Mobile terminal

3 Mobile Terminal

Figure 2 shows the picture of a mobile terminal. Figure 3 shows the structure of the mobile terminal. The mobile terminal consists of an Arduino board with sensors and actuators, and an Android terminal. The Arduino board and the Android terminal are connected with a USB cable and the Android Open Accessory Development Kit (ADK) [9]. In the Android terminal, the application program of the system is running. The application program consists of components such as Basic, AdkService, PukiwikiJava-ConnectorService (PJC-S), AdkThread, and AdkWiki-Activity. The AdkService is the central component of the mobile terminal and this is in charge of controlling other components. The AdkWikiActivity is in charge of GUI of the terminal. The PukiwikiJavaConnectorService is in charge of communicating with wiki sites. The Basic is the data processor of the M2M system. Packages such as android.jar, usb.jar, and maps.jar are used for the USB communication.

4 Data Processor

We have applied the programming environment of the reference [7] to this data processor. The language of the data processor is a basic-like programming language. Dimensions of the language are translated into hash tables just like the programming language Lua [15]. The data processor includes a parser and an interpreter. The parser inputs source programs or commands and the parser translates them into internal representation of programs or commands. The internal representation is a binary tree that can be evaluated by the interpreter. The interpreter is a Lisp interpreter that evaluates the internal representation of programs or commands, using the environment.

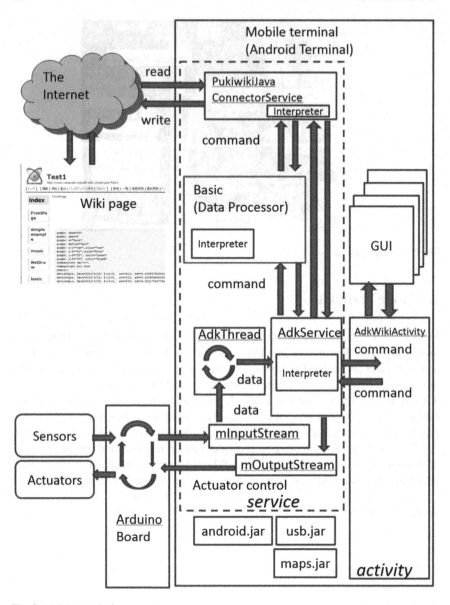

Fig. 3 Mobile terminal

The environment includes functions, variables and tables. The interpreter can send commands to command interpreters of the AdkService component and the Puki-wikiJavaConnectorService component. The AdkService component is in charge of communicating with sensors, actuators, GUI and the PukiwikiJavaConnectorService components. The Pukiwiki-JavaConnectorService is in charge of communicating with Pukiwiki sites.

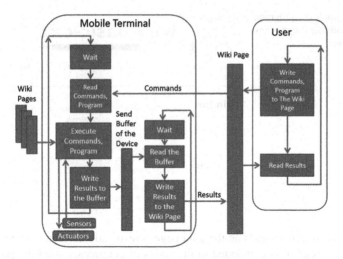

Fig. 4 Behavior of a mobile terminal

Fig. 5 GUI for giving the
initial URL

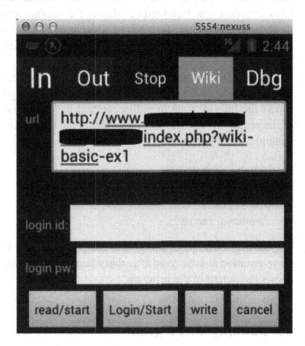

4.1 Reading the Program

Figure 4 shows behavior of a mobile terminal. The mobile terminal repeats to read
a wiki page whose initial URL is given by the GUI (Fig. 5) and after that, the URL
can be given by the command on a wiki page.

Fig. 6 Example of a program, which is embedded in a series of commands

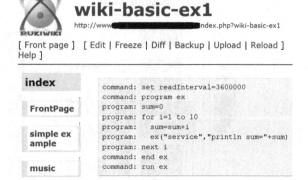

The series of commands on the wiki page is read and interpreted when the page has it. If a program is embedded in the series of commands, then the program is transferred to the parser of the Basic. The program is translated into its internal representation. The interpreter evaluates the internal representation of the program.

Figure 6 shows an example of a program, which is embedded in a series of commands. In this example, lines that start with "command:" are the commands. The first line,

$command : set\ readInterval = 3600000$ shows that the mobile terminal reads the page of the given URL every 1 h after that. The right hand side of the equation shows the interval time in milliseconds. Lines that start with "program:" are lines of the program. A program is enclosed by the following two command lines. In this example, these lines name "ex" the program.

    command: program ex
      command: end ex
    The last command line,
      command: run ex

for reading a wiki page shows that the program "ex" is translated into its internal representation and executed after that when this command line is interpreted. Figure 7 shows the output of the program.

4.2 Running the Program

A program of the Basic is a series of statements. After the program was translated into its internal representation, statements are executed one by one, from the first to the last. Some of statements of the program issue commands to command interpreters of the AdkService or PukiwikiJava-ConnectorService. They are represented as embedded functions. When a function of them was for inputting data from a wiki page or from a sensor of the terminal, the data is transferred to the interpreter as the

Fig. 7 The output of the
program of Fig. 6

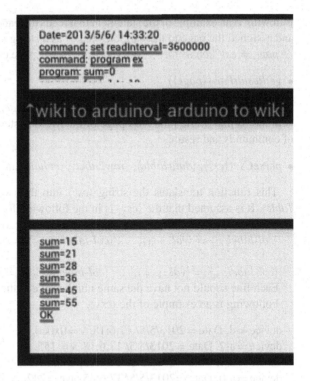

return value of the function. When a function of them was for outputting data to a
wiki page, an actuator, or the GUI, the data is transferred to them by an argument of
the command. The following line of the program in Fig. 4 shows the function that
outputs data to the GUI.

$$ex(\text{``service''}, \text{``println sum} = \text{''}+sum)$$

In this function ex, the first argument, *"service"*, represents the AdkService. The
second argument, *"printlnsum* $=''$ *+sum*, represents the command which is given
to the command interpreter of the AdkService. In this command, *"sum* $=''$ *+sum*, is
the argument of the command.

4.3 Some of Embedded Functions

The data processor includes following embedded functions.

- *ex(⟨object⟩, ⟨command⟩)*

As shown previously, this function sends the *<command>* in a string to the name
of the ⟨object⟩. Currently objects are "service" for the AdkService and "connector"
for the PukiwikiJava-ConnectorService. This function may have a return value. The

following is an example of the statement that reads the page of "http://www.page.ex/" and assigned the page to the variable "page" as a string value.

$page = ex($"connector"$, $"getpage$ http://www.page.ex/$")$

- getResultPart($\langle page \rangle$)

This function extracts the result part of the string $\langle page \rangle$. It is assumed that the page is in the format of the PukiWiki page of the M2M system that includes a sequence of commands and result.

- parseCsv($\langle csv \rangle$, $\langle dataTable \rangle$, $\langle rowLabel \rangle$, $\langle columnLabel \rangle$)

This function translates the string $\langle csv \rangle$ into the two dimensional array $\langle data Table \rangle$. It is assumed that the $\langle csv \rangle$ is in the following format.

$\langle col\text{-}label_{-1} \rangle = \langle val_{-1-1} \rangle, \ldots, \langle col\text{-}label_{-1-1n} \rangle = \langle val_{-1-1n} \rangle.$
$\langle col\text{-}label_{-2} \rangle = \langle val_{-2-1} \rangle, \ldots, \langle col\text{-}label_{-2-2n} \rangle = \langle val_{-2-2n} \rangle.$
. . .

$\langle col\text{-}label_{-m} \rangle = \langle val_{-1-m} \rangle, \ldots, \langle col\text{-}label_{-m-mn} \rangle = \langle val_{-m-mn} \rangle.$
Each line should not have the same number of equations.
Following is an example of the $\langle csv \rangle$.

device $=$ d, Date $=$ 2013/5/5/ 17:6:18, v $=$ 0x0c0.
device $=$ a-2, Date $=$ 2013/5/5/ 17:6:18, v $=$ 155.
device $=$ a-1, Date $=$ 2013/5/5/ 17:6:18, v $=$ 53.
device $=$ a-0, Date $=$ 2013/5/5/ 17:6:45, ave $=$ 242,..., dt $=$ 100.
device $=$ a-0, Date $=$ 2013/5/5/ 17:7:53, ave $=$ 242,..., dt $=$ 100.
. . .

$\langle rowLabel \rangle$ is the hash table which has key-value pairs of ("rowcol", "row") and ("maxIndex", maximum row index of the table). The ("rowcol","row") shows that this hash table includes row information such like the maximum index of the row. $\langle columnLabel \rangle$ is the hash table which has key-value pairs of ("rowcol","col"), ("maxIndex", maximum column index of the table), ($\langle col\text{-}label_{-1} \rangle$, column index of the label), ..., ($\langle col\text{-}label_{-max} \rangle$, column index of the label whose index is the maximum column value in the table).

- sumif($\langle dataTable \rangle$, $\langle hash\ table\ of\ row\ or\ column \rangle$, $\langle index_{-1} \rangle$, $\langle operator \rangle$, $\langle operand \rangle$, $\langle index_{-2} \rangle$)

This function sums values at $\langle index_{-2} \rangle$ of row or column if the condition, which is represented by the $\langle index_{-1} \rangle$, $\langle operator \rangle$ and $\langle operand \rangle$, is satisfied. If the $\langle hash\ table\ of\ row\ or\ column \rangle$ is the hash table of $\langle rowLabel \rangle$, the value of the sumif will be the following.

$$\sum_{i=0}^{rmax-1} \left\{ \begin{array}{l} dataTable\,[i, index_2]\,, \\ if\ operator\,(dataTabe\,[i, index_1]\,, operand) \end{array} \right\}$$

In this expression, *rmax* is the maximum index of the row.

If the ⟨*hash table of row or column*⟩ is the hash table of ⟨*colLabel*⟩, the value of the *sumif* will be the following.

$$
\sum_{j=0}^{cmax-1} \left\{ \begin{array}{l} dataTable\left[index_2, j\right], \\ if\ operator\left(dataTabe\left[index_1, j\right], operand\right) \end{array} \right\}
$$

In this expression, *cmax* is the maximum index of the column.

- *countif*(⟨*dataTable*⟩, ⟨*hash table of row or column*⟩, ⟨*index*$_{-1}$⟩, ⟨*operator*⟩, ⟨*operand*⟩)

This function counts the number of row or column if the condition, which is represented by the ⟨*index*$_{-1}$⟩, ⟨*operator*⟩ and ⟨*operand*⟩, is satisfied. If the ⟨hash table of row or column⟩ is the hash table of ⟨*rowLabel*⟨, the value of the *countif* will be the following.

$$
\sum_{i=0}^{rmax-1} \left\{ \begin{array}{l} 1, \\ if\ operator\left(dataTabe\left[i, index_1\right], operand\right) \end{array} \right\}
$$

If the ⟨*hash table of row or column*⟩ is the hash table of ⟨*colLabel*⟩, the value of the *countif* will be the following.

$$
\sum_{j=0}^{cmax-1} \left\{ \begin{array}{l} 1, \\ if\ operator\left(dataTabe\left[index_1, j\right], operand\right) \end{array} \right\}
$$

5 Usage Example

We are developing a remote room monitoring system that uses this M2M system. Figure 8 shows the outline of this system. We successfully acquired human activity data, light strength data and temperature data at a remote room and showed them on wiki pages by using the monitoring system. These data can be available at 24 wiki pages for every hour (Fig. 9). The data processor of this chapter realizes totaling up these data into every day data of the month.

Figure 10 shows a part of the program for the totaling up. It is assumed that there are 31 wiki pages for each day of a month and every wiki page of them has the same program, for running the program. This program computes average values of sensors of every hour and writes them after the line of "result:" of this page. This program also controls LEDs of the mobile terminal if some conditions were satisfied.

In the Fig. 10, the equation,

$$
page = ex(\text{"connector"}, \text{"getpage"} + url + i)
$$

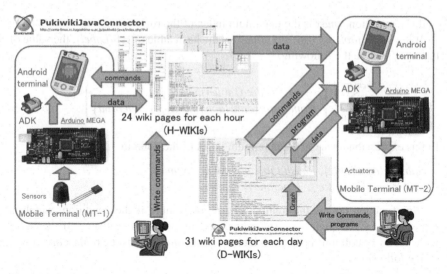

Fig. 8 Outline of the remote room sensor system

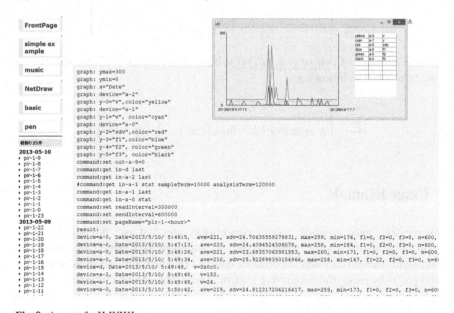

Fig. 9 A part of a H-WIKI page

gets the web page which has the sensor data of the hour of i.

The URL, which is the concatenation of *url* and i, represents the wiki page. The equation,

$$y0 = sumif(dataTable, rowLabel, columnLabel("device"),$$
$$" = ", "a - 0", columnLabel("sdv"))$$

Fig. 10 A part of the program
for totaling up 24 h pages into
a daily

```
command: set readInterval=3600000
command: program daily
program: dim dataTable
program: dim columnLabel
program: dim rowLabel
program: ex("service","clear sendBuffer")
program: output=""
program: url=http://www.*********.org/***/index.php?pir-1-
program: avemax0=0
program: avemax1=0
program: avemax2=0
program: avemin0=1024
program: avemin1=1024
program: avemin2=1024
program: for i=0 to 23
program:   page=ex("connector" , "getpage "+url+i)
program:   rpart=getResultPart(page)
program:   ex("service","println "+rpart)
program:   parseCsv(rpart,dataTable,rowLabel,columnLabel)
program:   y0=sumif(dataTable,rowLabel,columnLabel("device"),
program:       "=","a-0",columnLabel("sdv"))
program:   c0=countif(dataTable,rowLabel,columnLabel("device"),
program:       "=","a-0")
program:   ave0=1.0*y0/c0
program:   if ave0>avemax0 then avemax0=ave0
program:   if ave0<avemin0 then avemin0=ave0
program:   dataline="device=a-0, Date="+
program:       dataTable(0,columnLabel("Date"))+", ave="+ave0
program:   ex("service","println "+dataline)
program:   ex("service","putSendBuffer "+dataline)
         ...
program: next i
program: dataline="device=maxvalues, Data="+
program:     dataTable(0,columnLabel("Date"))
program: dataline=dataline+", avemax0="+avemax0+", avemax1="+avemax1+",
program:     avemax2="+avemax2
program: ex("service","println "+dataline)
program: ex("service","putSendBuffer "+dataline)
program: ex("service","sendResults.")
program: if avemax0 < 7.0 then ex("service","set out-a-8=255")
program:     else ex("service","set out-d-8=0")
program: if avemax1 < 100.0 then ex("service","set out-a-9=255")
program:     else ex("service","set out-d-9=0")
program: if avemax2 > 160.0 then ex("service","set out-a-10=255")
program:     else ex("service","set out-d-10=0")
command: end daily
command: set pageName="daily-1-<day>"
command: set result=""
command: run daily
result:
device=a-0, Date=2013/5/2/ 23:46:20, ave=5.650295889122299
device=a-1, Date=2013/5/2/ 23:46:20, ave=0.08333333333333333
device=a-2, Date=2013/5/2/ 23:46:20, ave=149.16666666666666
...
```

sums the values whose label is "sdv" in the CSV if the device of the sensor was
"a-0".

The equation,

$$c0 = countif(dataTable, rowLabel,$$
$$columnLabel(\text{"}device\text{"}), \text{"} = \text{"}, \text{"}a - 0\text{"})$$

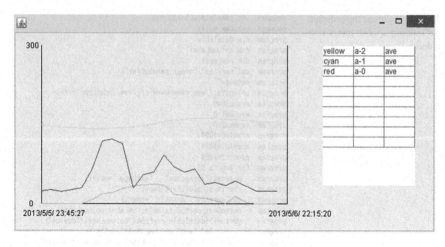

Fig. 11 Graph of the daily data

counts the lines of the device "a-0".
The equation,

$$ex(\text{"service"}, \text{"putSendBuffer"} + dataline)$$

adds the dataline to the "sendBuffer" of the AdkService and the equation,

$$ex(\text{"service"}, \text{"sendResults."})$$

writes the text on the buffer to the current page.
The equation,

$$ex(\text{"service"}, \text{"setout} - 8 - \langle val \rangle \text{"})$$

sets the value $\langle val \rangle$ to the actuator of "out-8".
The equation,

$$command: set\ pageName = \text{"daily} - 1 - \langle day \rangle \text{"}$$

changes the URL of the web page for next time reading. The $\langle day \rangle$ shows the current day.

Figure 11 shows the graph of the data of a day. The data was gained by totaling up the data of each hour of the day using the program of the Fig. 10. The lines of red, cyan and yellow show changes of human activity, light strength and temperature of the room respectively.

6 Comparisons with Related Work

6.1 IEEE 1888

IEEE 1888 [10] is a HTTP-based SOAP/XML over IP communication protocol among facilities, databases and information system for facility information management and control. A mobile terminal of our M2M system corresponding to the combination of the GW and the APP of IEEE 1888 system architecture. A Puki-wiki site of our M2M system is corresponding to the Storage of IEEE 1888 system architecture. Our M2M system does not have the Registry of IEEE 1888 system architecture. The program of a data processor of our M2M system is written on the wiki page of a wiki site whereas IEEE 1888 does not define where is the program of an APP.

6.2 Xively

Xively [11] is a real-time open data web service for the Internet of Things. As one of the most popular sensor data sharing services, Xively has open APIs for uploading and manipulating data, and so it is easy to enable a sensor device to upload data to Xively and also easy to write a program for processing the uploaded data. Many devices and applications can read and write data on Xively, and many users are currently using Xively.

Our M2M system has functions similar to those of Xively. However, the APIs of Xively are used for the Xively site only. In contrast, our system can be used for any PukiWiki site, not only a specific site. Our M2M system can also be used for building one's own sensor/actuator network.

6.3 Scripting Layer for Android

Scripting Layer for Android (SL4A) [12] brings popular scripting languages to Android. We use our original language processor for the data processor now. However the original language is not familiar with potential users. We should use SL4A instead of using our original language processor for such potential users of this M2M system.

6.4 Broadcast

Broadcast [13] is an embedded web application for remote Android device management. Broadcast uses the SL4A. However the Broadcast does not have the function for data exchanging between web pages.

6.5 Message Oriented Middleware

Message-oriented middleware (MOM) [14] is software or hardware infrastructure supporting sending and receiving messages between distributed systems. MOM allows application modules to be distributed over heterogeneous platforms and reduces the complexity of developing applications that span multiple operating systems and network protocols. Our data processor can be viewed as a MOM.

7 Concluding Remarks

We successfully acquired remote room data and totaled up the data using our M2M system. It is easy to define the data flow between wiki pages and it is easy to analyze the data of the M2M system by just writing programs on wiki pages. This easiness also can be a vulnerability of the security and safety. We are considering enhancing the security and safety of the M2M system. It is not so difficult to make the data processor not only for this M2M system but also for other purpose. We welcome the help of others who would like one to participate in improving and making his or her own sensor/actuator network.

References

1. Ward Cunningham: Wiki Wiki Web, http://c2.com/cgi/wiki?WikiWikiWeb. Accessed 7 May 2013
2. Wikipedia: http://www.wikipedia.org/. Accessed 7 May 2013
3. T. Yamanoue, K. Oda, K. Shimozono, A M2M system using Arduino, Android and Wki software, in *Proceedings of the 3rd IIAI International Conference on e-Service and Knowledge Management (IIAI ESKM 2012)*, Fukuoka, Japan, 20–22 Sept 2012, pp. 123–128
4. Android: http://www.android.com/. Accessed 7 May 2013
5. Arduino: http://www.Arduino.cc/. Accessed 7 May 2013
6. PukiWiki: http://pukiwiki.sourceforge.jp/. Accessed 7 May 2013
7. T. Yamanoue, K. Oda, K. Shimozono, A simple application program interface for saving java program data on a Wiki. Adv. Softw. Eng. **2012**, 9 (2012) (Hindawi Publishing Corporation)
8. T. Yamanoue, K. Oda, K. Shimozono, A malicious bot capturing system using a beneficial bot and Wiki. J. Inf. Process. (JIP), **21**(2), 237–245 (2013)
9. Android Open Accessory Development Kit: http://developer.android.com/tools/adk/. Accessed 7 May 2013
10. IEEE 1888: http://standardsinsight.com/engineering_standard_development/ieee1888. Accessed 23 May 2013
11. Xively: https://xively.com/. Accessed 23 May 2013
12. SL4A: http://code.google.com/p/android-scripting/. Accessed 7 May 2013
13. Broadcast: https://github.com/mleone/broadcast/. Accessed 7 May 2013
14. Sushant Goel, Hema Sharda, David Taniar: Message-Oriented-Middleware in a Distributed Environment, Proceeding of: Innovative Internet Community Systems, Third International Workshop, IICS 2003, Leipzig, Germany, June 19–21, 2003.
15. The Programming Language Lua: http://www.lua.org. Accessed 7 Dec 2013

Personal Ontology Extraction Considering Content Concordance from Tagging to Webpages in Similar SBM Users

Fumiko Harada and Hiromitsu Shimakawa

Abstract To realize web search engines with considering meaning of query phrases for each user, we have studied a method to extract hierarchical and synonymous relationships among tagged phrases on a social bookmark (SBM) for an individual SBM user. It detects the relationships from webpage clusters with same tagged phrases derived from the bookmarks shared in the target and his similar SBM users. However, noisy tagging violating personal phrase meaning degrades its detection accuracy. This chapter proposes a method to improve such drawback. The proposed method classifies webpages based on its content concordance as long as based on sameness of tagged phrases. Analyzing webpages belongingness to content-based and tag-based clusters, the relationships are detected more accurately. The experimental result shows the effectiveness of the proposed method.

Keywords Personal phrase meaning · Tagging · Social bookmark · Similar user

1 Introduction

WWW users want to find desirable many and only webpages from the vast number of webpages through web search engines. General web search engines generate search results based on correspondence between query phrases and the phrases in webpages. They do not consider *personal phrase meaning* assumed by each individual user. As an example, suppose that users A and B search by the query phrase "Web application". User A uses the phrase with implicit meaning "Wiki" and "BBS" while User B does with the meaning "Web mail" and "Web-based office software". A general search engine presents both of them the webpages including the text "Web application".

F. Harada (✉) · H. Shimakawa
Department of Information Science and Engineering, Ritsumeikan University 1-1-1,
Noji-Higashi, Kusatsu, Shiga 525-8577, Japan
e-mail: harada@is.ritsumei.ac.jp

R. Y. Lee (ed.), *Applied Computing and Information Technology*,
Studies in Computational Intelligence 553, DOI: 10.1007/978-3-319-05717-0_10,
© Springer International Publishing Switzerland 2014

User *A* receives the search result including the webpages about "Web mail" and "Web-based office software". User *B* similarly does those about "Wiki" and "BBS". The search result includes unnecessary webpages irrelevant to "Web application" in each user's mind.

In order to realize a search engine accounting for personal phrase meaning, personal phrase meaning must be identified by a formal description in advance. Thus, this research deals with identification of a personal phrase meaning by extracting a personal ontology defining hierarchical and synonymous relationships between phrases as the formal description. In the past work, the authors have proposed a method to extract a personal ontology from the phrases tagged to bookmarked webpages in a social bookmark (SBM) [1]. It detects the hierarchical and synonymous relationships among tagged phrases based on tag-based clusters of the webpages shared among the target and his similar users. However, it has a drawback that the detection accuracy becomes poor if there exist tagging violating the personal phrase meaning by tag omission, invalid tag selection, and label spelling error.

This chapter proposes an extension of the traditional method [1] to treat such invalid tagging. The proposed method classifies webpages based on concordance of contents as long as tagged phrases. Extending tag-based webpage clusters by content-based one, it eliminates shrink of tag-based clusters by invalid tagging and improves detection accuracy. The proposed method has the following characteristics:

- It does not extract general meaning of a phrase but personal meaning.
- It extracts personal ontology based on the webpages tagged with the phrase by the target and his similar SBM users.
- It extracts personal ontology against noisy information brought by invalid tagging.

We conducted an experiment to investigate the effectiveness of the proposed method over the traditional one. The experimental result shows that detection of hierarchical relationships by the proposed method improves the F-measure by up to 0.074 and improves the precision by up to 0.209 under the recall more than 0.2. For extraction of synonymous ones, the F-measure is improved by up to 0.0417 and the maximum precision is done by up to 0.218 under the recall more than 0.1. Moreover, the proposed method especially improves hierarchical relationship detection accuracy for the users who tend to give two tags with some relationship other than hierarchical one to an identical webpage. It also especially improves the synonymous relationship detection accuracy for the users who have enough number of bookmarks in every genre that he is interested in.

The rest of this chapter is organized as follows. Section 2 describes related work. Section 3 reviews the traditional method [1]. Section 4 proposes a novel method against the drawback of the traditional one. The experimental result is shown in Sect. 5. Section 6 concludes this chapter.

2 Related Work

A SBM, such as Hatena Bookmark [2] and Delicious [3], is instantiated as a tool which we can utilize for automatic extraction of phrase meaning. In a SBM, a user manages and organizes his bookmarks by freely tagging one or more phrases to them. Which phrase is tagged to which webpage is determined by him according to his desirable manner. It leads that his personal phrase meaning may be implied by the tagging to the bookmark. Analyzing tagged phrases from the viewpoint of tagged webpages will enable automatic estimation of his personal phrase meaning.

Some researchers have studied extraction of relationships between tagged phrases in a SBM [1, 4, 5]. In the method proposed in [4], relationships between tagged phrases are extracted from the data on user, resource and tags derived by using Probabilistic Latent Semantic Indexing. The research of [5] has proposed a method to represent relationships between tagged phrases by directed acyclic graph. These methods derive relationships from the viewpoint of the general tendency among entire SBM users.

As the method focusing on the phrases' relationships in personal mind, the authors have proposed a method to extract individual phrase meaning of a SBM user based on webpage clusters tagged with an identical phrase [1]. However, its detection accuracy becomes poor if a SBM user tags phrases to some webpages by the way unconsciously violating his phrase meaning. Improvement against such tagging is required for accurate extraction of the personal ontology.

3 Review of the Traditional Method

3.1 Relationship Between Phrases and Tagged Webpages

This subsection firstly describes the basic idea to extract personal ontology from SBM through an example, which is applied in the traditional method [1].

Suppose that each of SBM users A and B, whose tastes are similar, gives tags to the set of webpages $\{w_1, \ldots, w_{10}\}$ as shown in Table 1. For each tag, the set of webpages to which a user gives the tag is derived. we call it as the user's *tag cluster* for the tag. The tag clusters of Users A and B are derived as Fig. 1 from Table 1.

This research assumes that the tag cluster for a tag implies the personal meaning of the label phrase for the corresponding user. It is because an identical tag label is used in different meaning among users. For example, though both of Users A and B use the same tag label "Recommend" in Fig. 1, the corresponding tag clusters are completely different. It may be the case that User B's "Recommend" tag cluster, $\{w_1, \ldots, w_6\}$, are the webpages about information recommendation technology while that of User A, $\{w_7, \ldots, w_{10}\}$, have the contents about recommended products from online shops. In this case, the personal meaning of "Recommend" is information recommendation technology for User B while it is recommended products for User A. On the other

Table 1 An example of tagging by two users

Webpage	User A's tag	User B's tag
w_1	Information recommendation, collaborative filtering	Recommend
w_2	Information recommendation	Recommend
w_3	Information recommendation, collaborative filtering	Recommend
w_4	Information recommendation	Recommend
w_5	Information recommendation	Recommend
w_6	Information recommendation, collaborative filtering	Recommend
w_7	Recommend	Shop selection
w_8	Recommend	Shop selection
w_9	Recommend	Shop selection
w_{10}	Recommend	Shop selection

hand, "Recommend" tag cluster of User A is equal to "Shop selection" tag cluster of User B. These tag clusters shows that the meaning of "Recommend" for User A correspond with that of "Shop selection" for User B though the descriptions are different.

Hierarchical and synonymous relationships between two tagged phrases for a user will be extracted based on their tag clusters.

Two tags in a perticular relationship may be given to same webpages [6]. Comparing tag clusters of two tags in hierarchical one, one of the tag cluster may be the superset of the other, where the former is at the upper level and the latter is at the lower level. The former have broader meaning while the latter does narrower one. In Fig. 1, User A's "Information recommendation" tag cluster is superset of his "Collaborative filtering" tag cluster. In this case, the tag labels "Information recommendation" and "Collaborative filtering" are in hierarchical relationship where the former has the broader meaning and the latter does the narrower. In the user's personal phrase meaning, The concept of "Information recommendation" completely includes the concept of "Collaborative filtering".

Moreover, two tags in synonymous relationship may have a common tag cluster. In Fig. 1, User A's "Information recommendation" tag cluster is same with User B's "Recommend" one. The meaning of "Information recommendation" for User A and that of "Recommend" for User B have similar meaning. If personal phrase meaning of Users A and B are similar, the meaning of "Information recommendation" and "Recommend" for User A may be same, which means that they will be in synonymous relationship in User A's personal phrase meaning.

3.2 Overview of the Traditional Method

The traditional method [1] extracts personal ontology based on the idea given in the previous subsection.

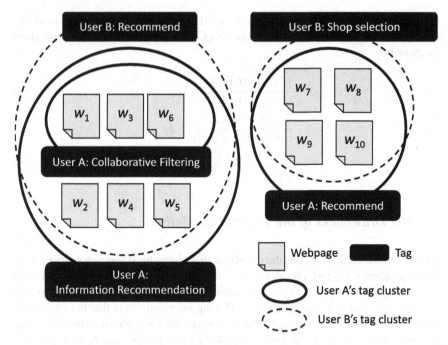

Fig. 1 An example of tag clusters and their relationships

Suppose tags T_x and T_y given by the target user. Denoted by $C(T_x)$ and $C(T_y)$ are his tag clusters of T_x and T_y, respectively. If all of the following equations hold, the hierarchical relationship where the label of T_x is at the upper level and that of T_y is at the lower is detected.

$$\frac{|C(T_x) \cap C(T_y)|}{|C(T_x) \cup C(T_y)|} \geq \theta_1, \tag{1}$$

$$\frac{|C(T_x) \cap C(T_y)|}{\min(C(T_x), C(T_y))} \geq \theta_2, \tag{2}$$

$$\frac{2|C(T_x) \cap C(T_y)|}{(|C(T_x)| + |C(T_y)|)} \geq \theta_3, \tag{3}$$

$$|C(T_x)| > |C(T_y)|, \tag{4}$$

where θ_1–θ_3 are given parameters. These equations mean $C(T_y)$ is almost contained by $C(T_x)$.

Suppose also that T_z is a tag given by the similar SBM user of the target user. The synonymous relationship between the label phrases of T_x and T_z may be detected from the fact that $C(T_x)$ and $C(T_z)$ are almost the same. It is reasonable since similar users may have similar classification of meanings of objects even if the words for calling the objects somewhat differ. Similar users in SBM can be identified through the method proposed in [7], for example. If all of the following equations hold, the synonymous relationship between the labels of T_x and T_y is detected. δ_1–δ_3 are given parameters.

$$\frac{|C(T_x) \cap C(T_y)|}{|C(T_x) \cup C(T_y)|} \geq \delta_1 \tag{5}$$

$$\frac{|C(T_x) \cap C(T_y)|}{\min(C(T_x), C(T_y))} \geq \delta_2 \tag{6}$$

$$\frac{2|C(T_x) \cap C(T_y)|}{(|C(T_x)| + |C(T_y)|)} \geq \delta_3 \tag{7}$$

3.3 The Drawbacks of the Traditional Method

The traditional method [1] has the drawback that hierarchical and synonymous relationships cannot be correctly detected if the target user tags phrases by the way disagreeing with his personal phrase meaning. Such tagging is *tag omission, invalid tag selection*, and *label spelling error*. The tag omission refers that the target user forgets to give a tag T_x to a webpage relevant to the label phrase. The invalid tag selection does that the target user incidentally gives another tag T_y to a webpage instead of the tag T_x which must be tagged originally. In this case, the user uses T_y for other webpages validly by the original meaning. T_y is invalid tag which must not be tagged to the webpage. The label spelling error refers the similar situation where a phrase T_z is tagged incidentally by spelling error instead of the phrase T_x which the target user wanted to tag in actual. Such invalid tagging shrink the tag cluster $C(T_x)$ compared with the one which must be generated in origin. The tag cluster $C(T_y)$ broadens out invalidly in contrast. The tag cluster $C(T_z)$ is generated as a set of only one webpage incidentally.

The traditional method [1] cannot eliminate the negative effect by these taggings, especially on the tag cluster $C(T_x)$ in the above-mentioned example. Shrink of $C(T_x)$ can be more significant and serious than broaden out of $C(T_y)$ and generation of $C(T_z)$. It may cause more often because T_x is tagged consciously by the target user while broaden out $C(T_y)$ and generation of $C(T_z)$ cause incidentally. Correct detection of hierarchical and synonymous relationships on T_x is prevented by this factor. The tag omission, invalid tag selection, and label spelling error by the target and similar SBM users are noisy information which disables accurate extraction of personal ontology. Thus, accurate extraction of personal ontology from SBM requires excluding the negative effect by the such noisy tagging.

4 Extraction of Personal Ontology of SBM User Considering Sameness of Webpage Contents

4.1 Method Overview

This section proposes a novel method by extending the traditional one [1] in order to eliminate shrink of tag clusters caused by noisy tagging. The proposed method utilizes not only tag clusters based on the sameness of the tags but also the *content clusters* from the sameness of the content of the webpages. A content clusters is defined as a set of bookmarked webpages whose contents are concordant. Extending tag cluster based on content clusters improves detection accuracy against shrink of tag clusters. The conception considers that even if two webpages are tagged with different label phrases, they should belong to a same tag cluster in actual if their contents are concordant. Including webpages with different tagged phrase but with the same content to a tag cluster, the size of the tag cluster is modified under guaranteeing invariant implied meaning to the corresponding phrase.

The proposed method extracts the personal ontology of a SBM user from his and his similar users' taggings to bookmarked webpages. The personal ontology represents the hierarchical and synonymous relationship in the personal phrase meaning of the target user, similar to the traditional method. The overview of the proposed method is shown in Fig. 2. Its procedure is as follows:

Step 1: Construction of tag clusters
Step 2: Extension of tag clusters based on content clusters
Step 3: Detection of hierarchical relationships based on extended tag clusters
Step 4: Detection of synonymous relationships based on extended tag clusters

The details of Steps 2–4 are discussed in Sects. 4.2, 4.3, and 4.4.

4.2 Extension of Tag Cluster

Step 2 in the previous subsection generates content clusters by classifying the bookmarked webpages based on concordance of contents. A nonhierarchical clustering algorithm can be applied to this classification. Then the tag clusters extracted in Step 1 is expanded based on the content clusters. If a tag cluster and a content cluster are similar, these clusters are regarded as the sets of webpages which should be tagged in actual with the phrase corresponding to the tag cluster. The union of these clusters is defined as the *extended tag cluster*.

Formally describing, suppose that the target user u_i and his similar user set SU_i share totally l bookmarks $\mathcal{R} = \{r_1, \ldots, r_l\}$. The tags given by u_i is denoted by $\{T_{i1}, \ldots, T_{im_i}\}$ and the label phrase of T_{ix} is $L(T_{ix})$. The tag cluster $C(T_{ix}) \subseteq \mathcal{R}$ is the set of webpages tagged with T_{ix}. Step 2 applies a nonhierarchical clustering to $\mathcal{R}$ and $\mathcal{R}$ is classified into a set of content clusters $\mathcal{G} := \{G_1, \ldots, G_a\}$, where

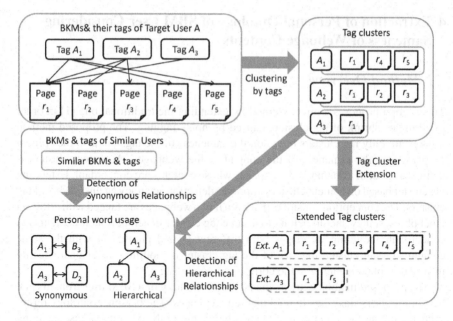

Fig. 2 Method overview

$G_j \cap G_k = \phi$ for $j \neq k$ and $\bigcup_{j=1}^{a} G_j = \mathcal{R}$. Suppose a content cluster set $\mathcal{G}' \subseteq \mathcal{G}$ where every content cluster G_j of $\mathcal{G}' \subseteq \mathcal{G}$ satisfies Eq. (8) for $C(T_{ix})$.

$$\frac{C(T_{ix}) \cap G_j}{C(T_{ix}) \cup G_j} \geq ET,   \tag{8}$$

where ET is a given parameter. The extended tag cluster $EC(T_{ix})$ is derived from $C(T_{ix})$ and $\mathcal{G}'$ by the following equation.

$$EC(T_{ix}) = \left(\bigcup_{g \in \mathcal{G}'} g \right) \cup C(T_{ix}),   \tag{9}$$

If no content cluster satisfies Eq. (8) on $C(T_{ix})$, the extended tag cluster $EC(T_{ix})$ is set to $C(T_{ix})$ without any change.

Extended tag clusters are also generated for each similar user $su_j \in \mathcal{SU}_i$.

4.3 *Extraction of Hierarchical Relationship*

The proposed method regards the extended tag cluster $EC(T_{ix})$ as the meaning of the phrase $L(T_{ix})$. If the size of $EC(T_{ix})$ is large, the personal meaning of $L(T_{ix})$ is supposed to be relatively broad for u_i. On the other hand, it may be relatively narrow if the size of $EC(T_{ix})$ is smaller.

Extending the traditional method, the hierarchical relationship with T_{ix} at the upper level and T_{iy} at the lower level is detected when all of the following equations hold:

$$\frac{|EC(T_{ix}) \cap EC(T_{iy})|}{|EC(T_{ix}) \cup EC(T_{iy})|} \geq \theta_1, \tag{10}$$

$$\frac{|EC(T_{ix}) \cap EC(T_{iy})|}{\min(EC(T_{ix}), EC(T_{iy}))} \geq \theta_2, \tag{11}$$

$$\frac{2|EC(T_{ix}) \cap EC(T_{iy})|}{(|EC(T_{ix})| + |EC(T_{iy})|)} \geq \theta_3, \tag{12}$$

$$|EC(T_{ix})| \geq |EC(T_{iy})|, \tag{13}$$

where θ_1, θ_2, and θ_3 are given thresholds.

4.4 *Extraction of Synonymous Relationship*

Suppose that the target user u_i gives the tags $\{T_{i1}, \ldots, T_{im_i}\}$. Denoted by $\mathcal{SU}_i$ is the similar users of u_i. For each user $su_j \in \mathcal{SU}_i$, his tags $\{T_{j1}, \ldots, T_{jm_j}\}$ are extracted. If the tag T_{ix} given by u_i and T_{jy} done by su_j have similar extended tag clusters, these label phrases $L(T_{ix})$ and $L(T_{jy})$ can be judged as synonyms. The synonymous relationship is detected by satisfying all of the following equations, where δ_1, δ_2, and δ_3 are given thresholds:

$$\frac{|EC(T_{ix}) \cap EC(T_{jy})|}{|EC(T_{ix}) \cup EC(T_{jy})|} \geq \delta_1 \tag{14}$$

$$\frac{|EC(T_{ix}) \cap EC(T_{jy})|}{\min(EC(T_{ix}), EC(T_{jy}))} \geq \delta_2 \tag{15}$$

$$\frac{2|EC(T_{ix}) \cap EC(T_{jy})|}{(|EC(T_{ix})| + |EC(T_{jy})|)} \geq \delta_3 \tag{16}$$

5 Experimental Evaluation

5.1 Experimental Contents

We conducted an experiment to verify the advantage of the proposed method compared with the traditional one [1].

The examinees were 4 university students, named u_1–u_4, who were Hatena Bookmark [2] users.

The reason of the smallness of the number of examinees is as follows. In actual SBM, since the strategies of webpage tagging are considerably different among users, it is considered that the proposed method will work significantly well in certain types of users while it will not work very well in other types of users. To analyze how the proposed method improves the traditional for which types of users in detail, we evaluated the effectiveness of the proposed method deeply on small number of examinees whom we could interview directly instead of evaluation among huge number of general on-line users.

In below, we abbreviate hierarchical and synonymous relationships to HR and SR, respectively, for simplicity of notation.

5.1.1 Generation of Correct Data by Questionnaire

We collected the data for verification through questionnaire to each examinee.

We firstly prepared the data of a similar user group for each examinee. We collected the bookmark data of the candidate similar users $SU = \{u_5, \ldots, u_{1059}\}$ from Hatena Bookmark, where SU were selected according to analysis of their bookmarks. For each $u_i \in \{u_1, \ldots, u_{1059}\}$, we collected his tagging information consisting of the tagged phrase $L(T_{ix})$ and tag cluster $C(T_{ix})$. We determined the similar user group $SU_i \subseteq SU$ for each examinee u_i ($i = 1, \ldots, 4$). SU_i is the set of $u_j \in SU$ who has at least one tag T_{jy} which satisfies $|C(T_{ix}) \cup C(T_{jy})| \geq 1$ for a tag T_{ix} of examinee u_i.

In the questionnaire on HR to each examinee u_i, each question presents a pair of the tagged phrases $(L(T_{ix}), L(T_{iy}))$ which satisfy $|C(T_{ix}) \cap C(T_{iy})| \geq 1$. T_{ix} and T_{iy} were selected randomly and partially from all pairs of his tags because number of the pairs was huge. He answered whether he considers that $L(T_{ix})$ and $L(T_{iy})$ have HR where $L(T_{ix})$ is upper level or not, and whether they have HR where $L(T_{ix})$ is lower level or not.

In each question of the questionnaire on SR to each examinee u_i, a pair of tagged phrases $(L(T_{ix}), L(T_{jy}))$ was presented. T_{jy} is a tag of a similar user $u_j \in SU_i$. Since the number of possible pairs was huge, we preferentially presented the pairs with high values of $|C(T_{ix}) \cap C(T_{jy})|$. He answered whether he considers $L(T_{ix})$ and $L(T_{jy})$ are synonyms or not.

The numbers of the presented tag pairs in the questionnaires are shown in Table 2. The examples of the presented tags and answers are also shown in Tables 3 and 4 for HR and SR questionnaires, respectively.

Table 2 The number of presented tag pairs in questionnaire

Examinee	Number of tag pairs	
	HR	SR
u_1	329	745
u_2	543	497
u_3	374	453
u_4	549	329

Table 3 Example of presented tags and answers in HR questionnaire

Tag pair		Answer	
T_{ix}	T_{iy}	HR[b] (T_{ix} at upper)	HR[c] (T_{iy} at upper)
Android	Windows	No	No
Apuri[a]	Windows	No	Yes
Puroguramingu[a]	Android	Yes	No

[a] In Japanese
[b] The HR existence where T_{ix} is upper level and T_{iy} is lower level
[c] The HR existence where T_{iy} is upper level and T_{ix} is lower level

Table 4 Example of presented tags and answers in SR questionnaire

Tag pair		Answer
T_{ix}	T_{iy}	SR existence
Detabesu[a]	Database	Yes
Dropbox	Web sabisu[a]	No
Chrome kakucho[a]	ado-on[a]	Yes

[a] In Japanese

We call the HRs and SRs obtained through the questionnaire as the *correct HRs* and *correct SRs*, respectively.

5.1.2 Generation of Evaluation Data

We detected HRs and SRs by each of the proposed and traditional methods for the pair of tag presented to each examinee in the questionnaire. The parameters $\theta_1–\theta_3$, $\delta_1–\delta_3$, and ET were set to the values from 0 to 1 by every 0.1. We used the Repeated Bisection method [8] for the nonhierarchical clustering in Step 2 of the proposed method. The number of clusters a in this algorithm was set from 1,000 to 310,000 by every 1,000 in HR and from 5,000 to 30,000 by every 5,000 in SR, respectively. The HR and SR detections were executed in all combination of the values of $\theta_1–\theta_3$, $\delta_1–\delta_3$, ET, and a. We call HRs and SRs detected by a method as the *detected HRs* and *detected SRs* by the method.

We evaluate the detection accuracy by the proposed/traditional methods based on the correspondence between the detected HRs and SRs, and the correct ones.

Table 5 HR: F-measure of detection

Examinee	Method	Precision	Recall	F-measure
u_1	Traditional	0.783	0.678	0.737
	Proposed	0.814	0.702	0.811
u_2	Traditional	0.323	0.485	0.388
	Proposed	0.317	0.600	0.415
u_3	Traditional	0.188	0.588	0.285
	Proposed	0.191	0.601	0.290
u_4	Traditional	0.801	0.683	0.737
	Proposed	0.721	0.783	0.751

Table 6 HR: Precision under given minimum recall

Examinee	Method	Minimum recall						
		0.1	0.2	0.3	0.4	0.5	0.6	0.7
u_1	Traditional	0.809	0.809	0.809	0.809	0.798	0.785	–
	Proposed	0.857	0.829	0.829	0.829	0.814	0.814	0.814
u_2	Traditional	0.295	0.295	0.295	0.295	0.295	–	–
	Proposed	0.505	0.505	0.447	0.364	0.324	0.317	–
u_3	Traditional	0.227	0.222	0.218	0.189	0.188	–	–
	Proposed	0.286	0.235	0.227	0.191	0.191	0.191	–
u_4	Traditional	0.854	0.854	0.854	0.854	0.854	0.819	0.759
	Proposed	0.974	0.946	0.931	0.927	0.862	0.830	0.770

5.2 Experimental Result and Evaluation

5.2.1 Hierarchical Relationship

Table 5 shows the detection accuracies on HR of the proposed and traditional methods. It shows the maximum F-measure among all combinations of parameters setting for θ_1–θ_3, δ_1–δ_3, ET, and a. Its corresponding precision and recall are also shown. It proves that the F-measure for every examinee by the proposed method exceeds that by the traditional one. The maximum improvement is 7.41 % of u_1. The proposed method can detect HR more accurately under balancing precision and recall.

Note however that, this research aims to improve the search performance of web search engines. If the proposed method can extract personal phrase meaning with high precision, the webpage ranking by search engines can be improved for individual user. Thus, we investigated the precision of the proposed method, however, under guaranteeing a certain value of the recall. Shown in Table 6 is the maximum precision among all of the parameter settings satisfying a given minimum recall. Each raw in "Minimum recall" represents the minimum recall to be guaranteed. For example, the maximum precision 0.857, the proposed method, and the minimum recall 0.1 for examinee u_1 means that the maximum precision of HR detection by the proposed method is 0.857 among all of the parameter settings which achieve equal or more than 0.1 recall. Note that "–" in the table is the case that no parameter setting achieve the corresponding minimum recall. Table 6 proves that the maximum precisions by the proposed method exceed those by the traditional one in all examinees and all minimum recall, which means that the proposed method improves HR detection accuracy in the case emphasizing precision. The maximum improvement is 0.209 for the minimum recalls 0.2 in u_2.

Finally, we evaluate the change of HR detection correctness for each correct HRs. It is desirable that the proposed method detects true-positively or true-negatively all HRs which are detected false-negatively or false-positively by the traditional one. The converse case is undesirable. We investigated how many desirable or undesirable changes of HR detection correctness occur. Moreover, we discuss for which types of users the proposed method improves the detection accuracy.

Table 7 shows the breakdown of the HR detection correctness by the proposed and traditional methods whose parameter settings are of the cases where the maximum precisions in Table 6 are achieved. It shows the numbers of true-positive, true-negative, false-positive, and false-negative detections by each method among all of the HRs in the tag pairs. The detection correctness of a HR is represented by [*the correctness by the traditional method:the correctness by the proposed method*], that is, [TP:TP], [TP:FN], [FN:TP], [FN:FN], [FP:FP], [FP:TN], [FN:TP], and [FN:FN]. TP, TN, FP, and FN represent true-positive, true-negative, false-positive, and false-negative detections, respectively. For example, [FN:TP] means correct HR of a tag pair is not detected by the traditional method while it is detected by the proposed method. [FN:TP] and [FP:TN] are desirable cases while [TP:FN] and [TN:FP] are undesirable ones. In examinees u_2–u_4, the cases of [FP:TN] contribute the improvement of detection accuracy. The proposed method improves the detection accuracy by mainly eliminating false-positive detection.

We discuss the characteristics of the tag pairs of [FP:TN] detections. In the traditional method, (T_{ix}, T_{iy}) satisfies $|C(T_{ix}) \cap C(T_{iy})| \geq 1$ from the viewpoint of being detected as having HR. Since the tags T_{ix} and T_{iy} are tagged at least one same pages, the phrases $L(T_{ix})$ and $L(T_{iy})$ have some association but HR. A tag pair which associate with each other but do not have HR may be the cause of [FP:TN] or [FP:FP] detections. Thus, we discuss more about the difference between [FP:TN] and [FP:FP] detections.

We consider why the proposed method improves false-positive HR detections. The proposed method generates extended tag clusters $EC(T_{ix})$ and $EC(T_{iy})$. Such process increases the size of tag clusters, that is, $|EC(T_{ix})| > |C(T_{ix})|$ and $|EC(T_{iy})| > |C(T_{iy})|$. It is accompanied by $|EC(T_{ix}) \cap EC(T_{iy})| \geq |C(T_{ix}) \cap C(T_{iy})|$. In the case that T_{ix} and T_{iy} do not have HR, the increment of $|C(T_{ix}) \cap C(T_{iy})|$ is smaller than that of $|C(T_{ix})|$ and $|C(T_{iy})|$. It means that $C(T_{ix})$ and $C(T_{iy})$ are extended by different content clusters, which is reasonable if T_{ix} and T_{iy} do not have HR. As the result, Eqs.(10)–(13) become not to hold for $EC(T_{ix})$ and $EC(T_{iy})$ even if they hold for $C(T_{ix})$ and $C(T_{iy})$. This leads the proposed method correctly judges anti-HR between T_{ix} and T_{iy} while the traditional one judges HR invalidly.

Investigating 1,650 HRs with [FP:FP] detection in detail, we found that $EC(T_{ix})$ and $EC(T_{iy})$ is larger than $C(T_{ix})$ and $C(T_{iy})$ for only 97 pairs. $C(T_{ix}) = EC(T_{ix})$ and

Table 7 HR: breakdown of detection correctness

Examinee	Correctness	Minimum recall						
	[Trad.:Prop.]	0.1	0.2	0.3	0.4	0.5	0.6	0.7
u_1	[TP:TP]	3	150	150	150	150	214	214
	[TP:FN]	0	8	8	8	8	1	1
	[FN:TP]	27	15	15	15	15	10	10
	[FN:FN]	262	119	119	119	119	67	67
	[FP:FP]	0	19	19	19	19	23	23
	[FP:TN]	1	1	1	1	1	0	0
	[TN:FP]	3	1	1	1	1	0	0
	[TN:TN]	33	16	16	16	16	14	14
u_2	[TP:TP]	61	61	74	106	132	140	–
	[TP:FN]	55	55	42	13	8	8	–
	[FN:TP]	0	0	1	0	1	0	–
	[FN:FN]	84	84	83	81	59	52	–
	[FP:FP]	48	48	67	133	183	223	–
	[FP:TN]	122	122	103	39	41	20	–
	[TN:FP]	0	0	3	0	6	2	–
	[TN:TN]	177	177	174	175	117	102	–
u_3	[TP:TP]	21	36	55	91	91	109	–
	[TP:FN]	40	5	6	3	3	0	–
	[FN:TP]	0	0	0	0	0	0	–
	[FN:FN]	87	107	87	54	54	39	–
	[FP:FP]	35	96	143	276	276	323	–
	[FP:TN]	120	12	12	6	6	1	–
	[TN:FP]	0	0	0	0	0	0	–
	[TN:TN]	240	287	240	113	113	71	–
u_4	[TP:TP]	58	82	104	136	179	228	276
	[TP:FN]	112	94	72	40	1	10	4
	[FN:TP]	5	7	7	8	0	1	2
	[FN:FN]	162	154	154	153	157	98	55
	[FP:FP]	3	0	3	6	17	22	30
	[FP:TN]	15	18	15	12	1	2	2
	[TN:FP]	0	0	0	0	0	0	0
	[TN:TN]	19	19	19	19	19	13	5

$C(T_{iy}) = EC(T_{iy})$ hold for the rest 1,553 pairs. It was also found that the reason of invariance between the original and extended tag clusters is because the content texts of the corresponding webpages were not validly extracted by automatic extraction. Removing such cases from the [FP:FP] cases, the precision is calculated as the rate of 522 [FP:TN] detections in $522 + 97 = 619$ [FP:*] ones, that is, 84.3 %. The proposed method therefore significantly improves false-positive detections. Because false-positive detections by the traditional method is caused by the tags which have some relationships but HR, the proposed one is especially effective for the users who tags to many webpages with two phrases which have some relationship but HR.

Table 8 SR: F-measure of detection

Examinee	Method	Precision	Recall	F-measure
u_1	Traditional	0.116	0.263	0.161
	Proposed	0.105	0.368	0.163
u_2	Traditional	0.205	0.844	0.330
	Proposed	0.219	0.711	0.335
u_3	Traditional	0.157	0.630	0.252
	Proposed	0.195	0.593	0.294
u_4	Traditional	0.171	0.476	0.252
	Proposed	0.177	0.476	0.258

Table 9 SR: Maximum precision under given minimum recall

Examinee	Method	Minimum recall								
		0.1	0.2	0.3	0.4	0.5	0.6	0.7	0.8	0.9
u_1	Traditional	0.125	0.116	0.075	0.075	0.070	0.070	0.070	0.061	–
	Proposed	0.125	0.116	0.105	0.081	0.081	0.070	0.076	0.061	–
u_2	Traditional	0.205	0.205	0.205	0.205	0.205	0.205	0.205	0.205	0.187
	Proposed	0.220	0.220	0.220	0.220	0.220	0.220	0.220	0.205	0.199
u_3	Traditional	0.211	0.157	0.157	0.157	0.157	0.157	0.146	–	–
	Proposed	0.286	0.286	0.195	0.195	0.195	0.183	0.162	0.148	0.137
u_4	Traditional	0.211	0.211	0.191	0.181	–	–	–	–	–
	Proposed	0.429	0.214	0.194	0.183	–	–	–	–	–

5.2.2 Synonymous Relationship

Shown in Table 8 is comparison of F-measures of SR detection between the proposed and traditional methods. Each F-measure is obtained as the maximum F-measure among all parameter settings of θ_1–θ_3, δ_1–δ_3, ET and a. The F-measures of the proposed method exceed those of the traditional one in all examinees. The maximum improvement is 0.0417 for u_3.

High precision is also preferable in SR detection in order to improvement of the ranking by web search engines. Table 9 provides the maximum precision under a given minimum recall. In Examinees u_2–u_4, the precisions are higher in the proposed method than in the traditional one for all minimum recalls. In u_1, though the precisions become poor for minimum recalls 0.1 and 0.2, it exceeds the traditional one's for more than 0.3 minimum recall. Similar to HR detection, the proposed method can improve SR detection performance from the viewpoint of emphasizing precision. The maximum improvement is 0.218 for minimum recall 0.1 in u_4.

We discuss for which types of tag pairs the proposed method works well in SR detection. Table 10 shows the breakdown of the SR detection correctnesss by the traditional and the proposed methods. [FP:TN] detection especially contributes the improvement from correctness by the traditional to the proposed. A remarkable fact is that the number of [TN:FP] detection is large in u_1 and u_4.

Table 10 SR: breakdown of detection correctness

Examinee	Correctness	Minimum recall								
	[Trad.:Prop.]	0.1	0.2	0.3	0.4	0.5	0.6	0.7	0.8	0.9
u_1	[TP:TP]	3	3	5	9	9	12	15	14	–
	[TP:FN]	0	0	0	0	0	0	0	0	–
	[FN:TP]	0	2	2	1	1	0	0	0	–
	[FN:FN]	16	14	12	9	9	7	4	5	–
	[FP:FP]	21	21	39	109	109	143	182	217	–
	[FP:TN]	0	0	2	8	8	19	22	0	–
	[TN:FP]	0	17	21	4	4	0	0	0	–
	[TN:TN]	561	544	520	461	461	420	378	365	–
u_2	[TP:TP]	27	27	27	27	27	27	27	38	38
	[TP:FN]	11	11	11	11	11	11	11	0	0
	[FN:TP]	0	0	0	0	0	0	0	0	6
	[FN:FN]	7	7	7	7	7	7	7	7	1
	[FP:FP]	96	96	96	96	96	96	96	147	169
	[FP:TN]	51	51	51	51	51	51	51	0	8
	[TN:FP]	0	0	0	0	0	0	0	0	8
	[TN:TN]	114	114	114	114	114	114	114	114	76
u_3	[TP:TP]	4	4	15	15	15	16	19	4	14
	[TP:FN]	0	0	1	1	1	1	0	0	0
	[FN:TP]	2	2	1	1	1	1	0	2	11
	[FN:FN]	21	21	10	10	10	9	8	21	2
	[FP:FP]	15	15	65	65	65	75	98	27	74
	[FP:TN]	0	0	21	21	21	25	13	2	6
	[TN:FP]	0	0	1	1	1	1	0	26	83
	[TN:TN]	290	290	218	218	218	204	194	250	142
u_4	[TP:TP]	0	0	13	17	–	–	–	–	–
	[TP:FN]	3	3	0	0	–	–	–	–	–
	[FN:TP]	3	3	0	0	–	–	–	–	–
	[FN:FN]	36	36	29	25	–	–	–	–	–
	[FP:FP]	0	0	54	77	–	–	–	–	–
	[FP:TN]	22	27	1	0	–	–	–	–	–
	[TN:FP]	4	11	0	0	–	–	–	–	–
	[TN:TN]	380	368	351	329	–	–	–	–	–

The factor of [TN:FP] detection of a tag pair (T_{ix}, T_{jy}) is discussed in below. Extension of $C(T_{ix})$ into $EC(T_{ix})$ increases the value of $|EC(T_{ix}) \cap EC(T_{jy})|$. [TN:FP] detection occurs in the case that the left-hand sides of Eqs. (14)–(16) increase invalidly on T_{ix} and T_{jy} which do not have SR. We investigated the reason of such invalid increment through additional questionnaire about the detailed relationship of the 57 tag pairs of [TN:FP] detection to Examinees u_1 and u_4. For 56 of the 57 tag pairs, the examinees answered the corresponding tags have HR. Moreover, in 42 of the 56 tag pairs, the tag cluster of the upper level tag had only one element. This causes that both of the tag clusters are extended by a same content cluster, which holds

Eqs. (14)–(16) invalidly. Poor number of bookmarked webpages corresponding to the upper level tag brings such situation. Thus, the proposed method works well in SR detection for the users and their similar ones who do not have a genre where enough relevant webpages are not collected in their bookmarks.

6 Conclusion

This chapter proposed a method to extract personal phrase meaning of a SBM user by analyzing his and his similar users' tagging to webpages. Though a traditional method [1] also has been studied for the same purpose, the proposed method in this chapter overcomes the false detection by the traditional one caused by noisy tagging. The proposed method generates content clusters based on the concordance of the contents of the bookmarked webpages as long as the tag clusters based on the sameness of the tagged phrases of the webpages adopted in the traditional method. Extending tag clusters based on content clusters, personal phrase meaning is validly detected against noisy tagging as hierarchical and synonymous relationships between tagged phrases.

The experimental result showed that the F-measure was improved by 0.0741 and the maximum precision under at least 0.2 recall was improved by 0.209 at most, for hierarchical relationships. For synonymous ones, the F-measure was done by 0.0417 and the maximum precision under at least 0.10 recall was improved by 0.218 at most. The proposed method could improve mainly false-positive detection into true-negative one. Moreover, the proposed method especially improved hierarchical relationship detection accuracy for the users who tends to give two tags with some relationship other than hierarchical relationship to an identical webpage. It also especially improves the synonymous relationship detection accuracy for the users who collects enough number of webpages in every genre in his boomark.

Acknowledgments Masaya Ito, who was a student of Graduate School of Science and Engineering, Ritsumeikan University until 2012, made enormous contribution to this research.

References

1. M. Ito, F. Harada, H. Shimakawa, Extracting ontology from tagging to web pages in similar user group. Int. J. Adv. Comput. Sci. **1**(2), 58–64 (2011)
2. Hatena Bookmark: http://b.hatena.ne.jp/
3. Delicious: http://delicious.com/
4. X. Wu, L. Zhang, Y. Yu, "Exploring Social Annotations for the Semantic Web", In WWW, pp. 417–426, May 2006
5. T. Eda, M. Yoshikawa, M. Yamamuro, Locally expandable allocation of folksonomy tags in a directed acyclic graph, In *Proceedings of the 9th international conference on Web Information Systems Engineering*, Lecture Notes in Computer Science, vol. 5175, pp. 151–162, 2008

6. S. Niwa, T. Doi, S. Honiden, Folksonomy tag organization method based on the tripartite graph analysis, In *Proceedings of IJCAI Workshop on Semantic Web for Collaborative Knowledge Acquisition*, Jan 2007
7. J. Diederich, T. Iofciu, Finding communities of practice from user profiles based on folksonomies, in *Proceedings of the 1st International Workshop on Building Technology Enhanced Learning Solutions for Communities of Practice*, 2006
8. Y. Zhao, G. Karypis, Comparison of Agglomerative and Partitional Document Clustering Algorithms, Technical report, Department of Computer Science, University of Minnesota, 2002

Psychophysiological and Behavioral Evaluation of the Process of Mastering Skills: To Select Appropriate Indices for a Target Movement

Hiroko Sawai, Kazune Tomotake, Yasuharu Ishii,
Keisuke Ueno and Emi Koyama

Abstract The purpose of this study was to evaluate the early process of mastering skills to reveal skill factors including pause and adjustment in producing traditional handicrafts using psychophysiological and behavioral indices. The indices were measured in two experiments. Tasks that needed obtaining skill factors were performed. As the results, heart rates, time-series behaviors of elecrooculogram (EOG), and wrist activities during tasks, and subjective scores before tasks reflected the difference of the product quality in the early process of mastering skills. From the series of studies, skill factors could be evaluated quantitatively using different indices appropriate for the target movement. Therefore, it is important to select appropriate indices for a target movement in practical use. To reveal skill factors quantitatively can support to archive and inherit the skills for producing traditional handicrafts.

Keywords Mastering skills · Pause · Adjustment · Psychophysiological evaluation · Behavioral evaluation

1 Introduction

Producing high-quality goods requires mastering skills to produce. It is not enough only to follow a procedure to achieve high quality. It needs to obtain skill factors as a part of mastering skills. However, by reason that the skill factors include implicit knowledge, it is difficult to understand and obtain. Especially, it takes many years for novices to master the skills for producing traditional handicrafts that requires high-level skills. In addition, aging and lack of successors in the industry of traditional handicrafts are social issues. Thus, revealing the skill factors quantitatively is required

H. Sawai (✉) · K. Tomotake · Y. Ishii · K. Ueno · E. Koyama
Graduate School of Science and Technology, Kyoto Institute of Technology, Kyoto, Japan
e-mail: hsawai@kit.ac.jp

R. Y. Lee (ed.), *Applied Computing and Information Technology*,
Studies in Computational Intelligence 553, DOI: 10.1007/978-3-319-05717-0_11,
© Springer International Publishing Switzerland 2014

Fig. 1 Conceptual diagram of a relationship between performance and psychophysiological states (Modified from [5])

to archive and inherit the skills for producing traditional handicrafts. It is supposed to support obtaining the skills effectively and preventing the extinction of the skills.

Several studies have focused on revealing skill factors for producing traditional handicrafts. The results of these studies showed that masters had specific rhythms measured by electromyogram signals of arms [1], accelerations of fingers [2], eye motions [3], and respirations [4] mainly compared to novices. It is suggested that the rhythms such as pause and adjustment are important elements of the skill factors.

On the other hand, performance such as exerting these rhythms is based on psychophysiological states. Figure 1 shows a conceptual diagram [5]. Decreased alertness causes lowering attentiveness such as short-term memory and psychomotor vigilance [6]. Lowering attentiveness causes errors and lowering performance. Therefore, preparing the optimal psychophysiological states is important to achieve high quality. Thus, both of psychophysiological states and behaviors should be evaluated to reveal skill factors quantitatively. However, it is not clear how psychophysiological states and behaviors are involved with product qualities.

In previous studies, the authors revealed the skill factors including pause and adjustment related to product qualities could be evaluated by time-series behaviors of electrooculogram (EOG) that might reflect psychophysiological states and behaviors [7]. Moreover, it was revealed that wrist activities, heart rates, near-infrared spectroscopy (NIRS) signals, and subjective scores could be indices for psychophysiological states or behaviors reflecting product qualities [8], but it suggested that selecting an appropriate index for the target movement is important for practical use [8]. In this study, selecting method of an appropriate index for a target movement and evaluation using the index for practical use were discussed with reconstructed previous studies.

The purpose in the series of studies was to evaluate the early process of mastering skills to reveal skill factors including pause and adjustment in producing traditional handicrafts using psychophysiological and behavioral indices. Especially the skills related to quality enhancement of products were focused. Psychophysiological and behavioral indices were analyzed in the following two experiments, in which tasks with movements involving different body parts were performed. Experiment A was performed to evaluate skills using hands and fingers by psychophysiological and behavioral indices in the process of mastering skills. Experiment B was performed to evaluate skills using arms by time-series behaviors of EOG by reason that eye movements interacted with hand movements [9, 10] and EOG might reflect both psychophysiological states [11, 12] and behaviors in the process of mastering skills.

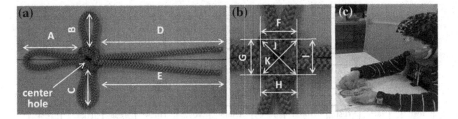

Fig. 2 a Sample figure, b sample figure of center hole and c experimental scene

2 Experiment A: Evaluation of Skills Using Hands and Fingers

2.1 Methods

2.1.1 Participants

Seven healthy paid volunteers (male, 21–22 years) participated in the experiments. All participants were instructed to keep a sleep-wake schedule (bedtime; 1:00 a.m., wake-up time; 8:00 a.m.) from 7 days before the experiments to the end of the experiments. Each participant was given an explanation about the research purpose and its procedures, and written informed consent was obtained. The protocol was approved by the ethics committee of Kyoto Institute of Technology.

2.1.2 Tasks

Tying a string to produce a traditional decoration was performed as a task requiring finger dexterous manipulation [13]. Figure 2a, b shows a sample of tied string and Fig. 2c shows the experimental scene. Participants were instructed to imitate the sample figure accurately with a procedure manual. This task was chosen as a task that can be evaluated the quality by measuring a size difference from the sample and does not bring movements making noises on biological signals.

2.1.3 Procedure

The experimental protocol consisted of 1-day practice, 4-day experiments in the laboratory, and 3-day training at home (Fig. 3). The tasks and subjective scoring were repeated 3 times a day from 10:00 to 12:00 a.m. A verbal fluency task was conducted to control psychophysiological state and confirm the condition at the beginning of the experiments. A task included 5 trials to tie a string. Each trial was tying one string within 45 s and resting 20 s before and after a trial (Fig. 3). This protocol was set to measure signals accurately by near-infrared spectroscopy (NIRS). Participants were

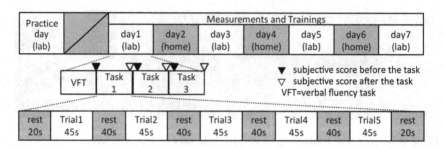

Fig. 3 Protocol; Experiment A

instructed to put their chin on the chin rest during the task, to grade the own trial score of tying a string (completed or not), and to evaluate subjective scores before and after each task. In addition, participants practiced 10 times a day for 3-training days at home.

2.1.4 Measurements and Data Analysis

Electroencephalogram (EEG), RR interval (RRI) of heart rates with electrocardiogram (ECG), NIRS signals (cortical oxygenated-hemoglobin concentration, Shimadzu Corp., FOIRE-3000), skin blood flow (OMEGAWAVE, Inc., FLO-C1), subjective scores with Visual Analog Scale (questions about psychophysiological states and the task), wrist activities (both of wrists, ActiGraph, ActiSleep Monitor), and task scores were measured (Fig. 4). Biological signals were recorded by Polymate AP1000 (TEAC Corp.) with active electrodes and analyzed by Bio Signal Viewer (NoruPro Light Systems, Inc.). The task scores were the achieved steps of procedure as an index of quantity (Table 1) and the relative size difference from the sample as an index of quality (the relative size difference = sum | each part size of a task product/each part size of the sample −1|, each part means A–K of Fig. 2a, b). Seven participants were analyzed. To control for a high degree of interindividual variability, the data of RRI was transformed to deviation from the 20 s pre-rest period, then normalized for each participant (RRI variation = (mean of RRI deviation during a trial-mean of RRI deviation during all trials)/standard deviation of that during all trials).

NIRS signals in motor area (Cz, International 10–20 system) were normalized based on the 20 s pre-rest period for each trial (NIRS signal = (mean of oxygenated-hemoglobin concentration during a task-mean of that during pre-rest)/standard deviation of that during pre-rest).

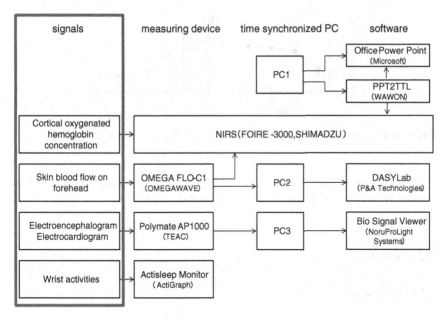

Fig. 4 Diagram of measurements; Experiment A

Table 1 The task score of achieved steps

	Index	Score
Not completed	No figure	0
	One loop is made	0.2
	Two loops are made	0.4
	One loop crosses the center hole	0.6
	Two loops cross the center hole	0.8
	Figure is completed but the participant grades it is not completed	0.9
Completed	No center hole	1.0
	Center hole is formed	1.2

2.1.5 Statistics

IBM SPSS Statistics 20 was used for all analyses. The task scores of achieved steps were analyzed by Friedman test. Post-hoc analysis was Wilcoxon signed-rank test using Bonferroni's procedure. The task scores of size difference were analyzed by repeated one-way analysis of variance (rANOVA). All P-values derived from rANOVA were based on Huynh-Feldt's corrected degree of freedom. Post-hoc analysis was paired-t test using Bonferroni's procedure. Correlations of RRI, wrist activities, NIRS signals, and task scores of size difference were calculated using Pearson's product-moment correlation coefficient. Values of each trial or task, instead of the

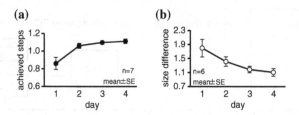

Fig. 5 Task scores **a** achieved steps, **b** size difference (relative score), modified from [8]

mean values in all participants, were used in the correlation analysis taking into account that each participant obtains the skills at different pace.

2.2 Results

2.2.1 Task Scores

The scores of achieved steps showed a trend to increase then become stable (Fig. 5a), and the scores of size difference decreased (Fig. 5b) through 4 days [8]. A main effect was observed for days in the scores of achieved steps (p = 0.006, Fig. 5a) and size difference of all participants through 4 days ($F_{3,15} = 7.841, p = 0.018$, Fig. 5b). The task scores had no significant differences between days. One participant could not complete tying a string at first day, and he was excluded from the analysis of size difference. Therefore, participants were in the process of quantitative and qualitative improvements of the skill to tie a string in 4 days.

2.2.2 Classification of the Process

To focus on the process of mastering skills for quality enhancement, each trial was classified whether tying a string was completed or not in all participants. The group with completed trials was classified as a quality group. The quality group included 332 trials in 7 participants out of total 420 trials, and 40 tasks in 6 participants out of total 84 tasks.

2.2.3 Relationship Between Indices

The scores of size difference in the quality group had correlations with the predictions of smooth movements (n = 38; There were two missing values, r = −0.525, p = 0.001, Fig. 6), the RRI variation (n = 332, r = −0.188, p = 0.001, Fig. 7a), and the wrist activities of both hands (n = 332, r = 0.110, p = 0.045, Fig. 7b). There

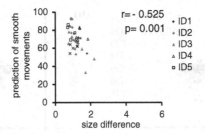

Fig. 6 Correlations between predictions of smooth movements and the scores of size difference, modified from [8]

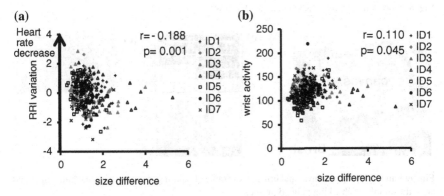

Fig. 7 Correlations with the scores of size difference **a** RRI, **b** wrist activity [8]

were no significant correlations in subjective scores related to psychophysiological states. The results of subjective scores, heart rates, and wrist activities showed the characteristics of psychophysiological states or behaviors related to product qualities.

The NIRS signals had no correlation with the scores of size difference in the quality group. However, the NIRS signals had a correlation with the RRI variation ($n = 291$; There were 41 missing values for the reason of measurement strain, $r = 0.208$, $p < 0.001$, Fig. 8a) and the wrist activities of both hands ($n = 291$, $r = -0.179$, $p = 0.002$, Fig. 8b) without a correlation with skin blood flow at forehead.

3 Experiment B: Evaluation of Skills Using Arms

3.1 Methods

3.1.1 Participants

Eight healthy paid volunteers (male, 20–26 years) participated in the experiments. All participants were instructed to keep the sleep-wake schedule (bedtime; 24:00,

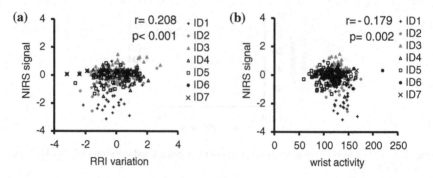

Fig. 8 Correlations with NIRS signals **a** RRI, **b** wrist activity [8]

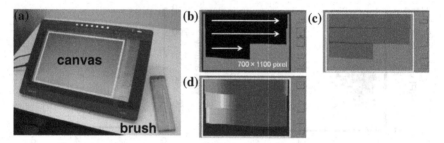

Fig. 9 Painting device **a** canvas and brush, **b** example of painting, **c** example of checking unpainted area, **d** example of checking painted pressure

wake-up time; 7:00 a.m.) from 7 days before the experiments to the end of the experiments. Each participant was given an explanation about the research purpose and its procedures, and written informed consent was obtained. The protocol was approved by the ethics committee of Kyoto Institute of Technology.

3.1.2 Tasks

Imitated painting to produce a traditional lacquer ware was performed as a task requiring steady movements of the arm [14]. Figure 9 shows the device for painting. A display (700 by 1,100 pixels, approximately 190 by 290 mm, Wacom DTU-1631C) and an input device formed a 50 mm-wide brush were used. One canvas can be painted by 4 strokes. Participants were instructed to paint from the left upper corner to right, maintain uniform pressure of painting, reduce unpainted area, and paint many canvases as possible. The results of painting pressures (color scale) and unpainted areas (grayscale) were displayed after the task to master the skills efficiently. This task was chosen as a task that can be evaluated the quality by measuring a uniformity of painting pressures and needs basic side to side arm movements that are common in some traditional skills.

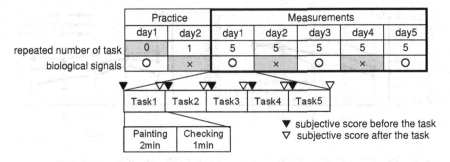

Fig. 10 Protocol; Experiment B

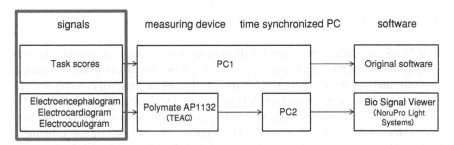

Fig. 11 Diagram of measurements; Experiment B

3.1.3 Procedure

The experimental protocol consisted of 2-day practices, 3-day experiments, and 2-day trainings (Fig. 10). The tasks and subjective scoring were repeated 5 times a day from 10:00 to 12:30. Each task included 2 min painting and 1 min checking the task scores. Painting a canvas was repeated for 2 min.

3.1.4 Measurements and Data Analysis

EEG, EOG (vertical and obliquely horizontal positions), ECG, subjective scores with Visual Analog Scale (questions about psychophysiological states and the task), and task scores were measured (Fig. 11). Task scores were the painted pixels for 2 min as an index of quantity and the variance of painted pressure for each canvas as an index of quality. The variance of painted pressure was calculated relative values from 500 levels that the display could detect. Seven participants were analyzed except for one participant (ID6) whose eye movements were different from others because of the different arm movements. EOG sampling rate was 500 Hz, the signals were digitally high-cut filtered at 30 Hz, and notch filter was set at 60 Hz. The sampling rate of EOG was converted to 100 Hz.

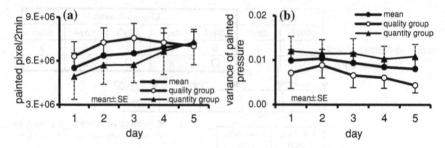

Fig. 12 Task scores **a** painted pixels, **b** variance of painted pressure (relative score) [8]

3.1.5 Statistics

IBM SPSS Statistics 20 was used for all analyses. Evaluation of systematic changes in task scores, EOG signals according to days was carried out using a repeated one-way analysis of variance (rANOVA) during the task periods. All P-values derived from rANOVA were based on Huynh-Feldt's corrected degree of freedom. Post-hoc analysis using Bonferroni's procedure and a paired-t test were performed. The auto-correlation coefficients and cross-correlation coefficients of EOG signals (obliquely horizontal positions) per canvas were calculated. The first peak and the lag were calculated with autocorrelation coefficients. The cross-correlation coefficient was calculated based on a stable signal that was the maximum of the first peak of the autocorrelation coefficient at task 5 on each day. The correlations between EOG signals and task scores were calculated using Pearson's product-moment correlation coefficient.

3.2 Results

3.2.1 Task Scores

The scores of painted pixels showed a trend to increase (Fig. 12a) [7, 8]. A main effect was observed for days in the scores of painted pixels of 7 participants through 5 days ($F_{4,24} = 3.099, p = 0.044$). The variance of painted pressure (Fig. 12b) and the task scores between days had no significant change. Therefore, participants were in the process of quantitative improvements of the skill to paint a canvas in 5 days.

3.2.2 Classification of the Process

To focus on the process of mastering skills for quality enhancement, participants were classified whether the scores of painted pixels were stable or not through 5 days. The classification was performed by each participant not by trial because of the difficulty

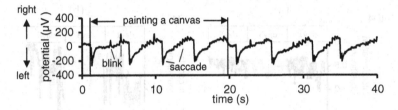

Fig. 13 Example of time-series behaviors of EOG, modified from [8]

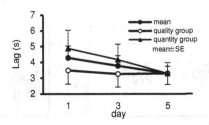

Fig. 14 The lags of autocorrelation coefficients of the first peaks, modified from [7]

to clear the stable level of the painting speed through experiments. The score of painted pixels also showed a trend to increase in 4 participants (quantity group: ID2, 3, 5, 7), and stable in 3 participants (quality group: ID1, 4, 8, Fig. 12a). The variance of painted pressure in quality group showed a trend to decrease (Fig. 12b).

3.2.3 Time-Series Behaviors of EOG

Figure 13 shows an example of time-series behaviors of EOG during painting canvases. EOG signals changed with the side to side arm movements, and it included signals of saccades and blinks.

As the quantitative trends through days, the lags of autocorrelation coefficients of the first peak showed a trend to decrease (Fig. 14). A main effect was observed for days in the lag of autocorrelation coefficient of 7 participants through 3-measurement days ($F_{2,12} = 7.699$, $p = 0.015$). This result indicated that the frequency of EOG rhythms associated with arm movements became shorter according to the increase of painting speed.

As the qualitative trends through days, the correlation coefficients had no common changes in quality group and quantity group. However, 3 participants (ID2, 4, 7) showed a trend to increase the autocorrelation and cross-correlation coefficient through days, and others (ID1, 3, 5, 8) showed unstable changes (Fig. 15a, b). Therefore, the 3 participants were in the process of improvement for EOG stability.

Focusing on the relationships between quality enhancement and indices, the variance of painted pressure had correlations with the autocorrelation coefficients of the first peaks (n = 387; 3 participants painted total 387 canvases, r = −0.237,

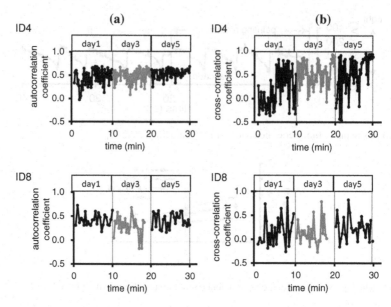

Fig. 15 Examples of correlation coefficient **a** autocorrelation coefficient; modified from [8], **b** cross-correlation coefficient; modified from [7], each plot shows correlation coefficient of EOG during painting a canvas

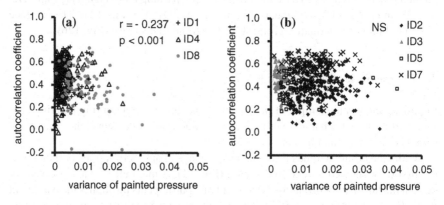

Fig. 16 Relationship between autocorrelation coefficient and variance of painted pressure **a** quality group, **b** quantity group, modified from [8]

$p < 0.001$, Fig. 16a) in the quality group (for reference, Fig. 16b). In addition, the variance of painted pressure had correlations with the cross-correlation coefficients ($n = 378$; baseline signals of cross-correlation were excluded from total canvases, $r = -0.133, p = 0.01$) in the quality group, and the trends were same as that of autocorrelation coefficients (Figure omitted).

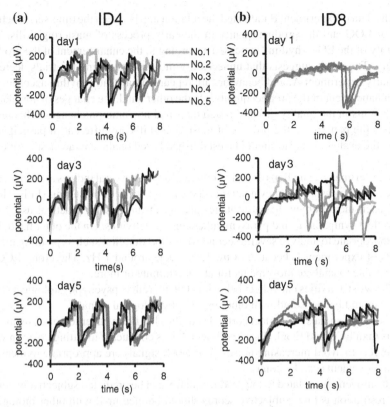

Fig. 17 Examples of time-series behaviors of EOG during painting first 5 canvases in each day, **a** ID4, **b** ID8 (ID8 painted 4 canvases on day1), modified from [8]

Focusing on EOG of ID4 as the most skilled participant by reason that the participant had good quantitative and qualitative task scores, the characteristics of the EOG were short cycles, fast movements from left to right, and constant accelerations, especially in day 5, compared to other participants (Fig. 17a, b).

4 Discussion

The data in experiment A indicated the relationships between behaviors and product quality, and also between psychophysiological states and the product quality in the early process of mastering skills. The wrist activities and heart rates decreased according to the enhancement of the product quality. The smooth movements related to the product quality could be predicted subjectively. Additionally, the NIRS signals increased according to the decrease of wrist activities and also heart rates. These results suggest that unnecessary movements and unnecessary sympathetic nerve activities during the trials decrease and cortical activities increase according to the enhancement of the product quality.

The data in experiment B indicated the relationship between the time-series behaviors of EOG and the product quality in the early process of mastering skills. The stability of the EOG rhythms increased according to the enhancement of the product quality. The result suggests that unnecessary movements during the trials decrease similarly experiment A, and behaviors are repeated in a steady rhythm according to the enhancement of the product quality. On the other hand, the changes of correlation coefficients in the quality group through days had no common changes. This result implies the patterns of the process of mastering skills are different in participants. Thus, the evaluation scale should be established based on the characteristics of each person.

In experiment A and B, psychophysiological and behavioral indices and subjective scores were measured and evaluated in the early process of mastering skills. However, each measurement has advantages and disadvantages for practical use. ECG can reflect sympathetic and parasympathetic nerve activities. On the other hand, it is not appropriate to analyze several seconds or a short duration such as painting movements of experiment B because of the ECG cycle around 1 Hz. Therefore, ECG is appropriate to analyze movements for about a minute or more.

The wrist activities can reflect behaviors but not reflect psychophysiological state, and it should be measured with other psychophysiological signals.

The NIRS signals can reflect cortical activities but it is weaker to noises by movements than ECG and EOG. The experiment A was conducted in sitting position with a chin rest to avoid increasing noises. The NIRS signals are appropriate to measure when movements can be controlled.

The movements related to the product quality can be predicted subjectively but its time resolution is low. Subjective scores should be measured with other biological signals when focusing on time-series movements.

EOG can reflect psychophysiological states and behaviors with higher time resolution than ECG, NIRS, and subjective scores. On the other hand, it needs repetition of movements to evaluate, and it is not appropriate to evaluate behaviors when the target movements involve with almost no eye movements such as movements with fingertips or without eye movements. Additionally, by calibrating the angle of eye movements and the angle or the position of head shifts, the accuracy of the analysis method is improved.

Focusing on the evaluation of the skill factors including pause and adjustment, it can be extracted as common characteristics by analyzing the velocity and acceleration of EOG waveforms in the process of mastering skills. In addition, EOG varies with movements, and includes psychophysiological signals such as saccades influenced by attention [11] and blinks influenced by cognitive states [12]. Therefore, it is supposed that EOG is a useful index to evaluate both psychophysiological states and behaviors. However, when targeting long time movements, the movement should be divided into short time units. Then each unit should be analyzed to keep the accuracy of this analysis method in practical use.

From the series of studies, skill factors could be evaluated quantitatively using different indices appropriate for a target movement. A summary about characteristics of time-series indices except subjective scores because of the discrete data is shown in

Table 2 Summary about characteristics of time-series indices

	Item	Wrist activities	NIRS signals	Heart rates	EOG
Duration of	Less than about 1 min	*	*	/	*
movements	About 1 min or more	*	*	*	*
Kinds of	With almost no eye movements	*	*	*	/
movements	With eye movements	*	*	*	*
	With head movement	*	/	*	(*)
Reflect	Psychophysiological states	/	*	*	*
characteristics	Behaviors	*	/	/	*

* appropriate for the item
/ not appropriate for the item
( ) recommended to measure optional index simultaneously

Table 2. These findings help to obtain skills effectively by using the index as feedback information or a sample, and to prevent the extinction of the skills by archiving the skill.

5 Conclusion

The early process of mastering skills including pause and adjustment were evaluated to reveal skill factors in producing traditional handicrafts using psychophysiological and behavioral indices. From the series of studies, skill factors could be evaluated quantitatively using different indices appropriate for a target movement. Therefore, it is important to select appropriate indices for the target movement in practical use. To reveal skill factors quantitatively can support to archive and inherit the skills for producing traditional handicrafts.

Acknowledgments This work was supported by Grant-in-Aid for Scientific Research (B), Grant Number 23300037.

References

1. H. Kato, K. Morimoto, *Examination of Skill for Sharpening Japanese Cooking Knives Based on Electromyogram*. Dentoumiraidayori '06. pp. 61–64 (2006) (in Japanese)
2. T. Tanaka, A. Ohnishi, M. Kume, M. Shirato, K. Tsuji, A. Nakai, T. Yoshida, *Interval of Weaving Kanaami*. Dentoumiraidayori '08. pp. 62–66 (2008) (in Japanese)
3. M. Iue, T. Ota, K. Hamasaki, M. Kume, Y. Yoshida, A. Nakai, N. Sasaoka, A. Goto, *Eye Motion Analysis in Chado, the Way of Tea*. Dentoumiraidayori '09. pp. 221–227 (2009) (in Japanese)
4. R. Ito, A. Hiyama, H. Namiki, M. Miyashita, T. Tanikawa, M. Miyasako, M. Hirose, Extraction of artisans' implicit knowledge for skill training. J. Inst. Image Inform. TV. Eng. **33**(21), 123–127 (2009) (in Japanese)

5. M. Moore-ede, *The Twenty Four Hour Society* (Addison-Wesley Publishing Company, Boston, 1992), pp. 44–63
6. J. Wyatt, A. Cecco, C. Czeisler, D. Dijk, Circadian temperature and melatonin rhythms, sleep, and neurobehavioral function in humans living on a 20-h day. Am. J. Physiol. **277**, R1152–R1163 (1999)
7. H. Sawai, K. Tomotake, K. Ueno, E. Koyama, in A study of quantitative evaluation about the process of mastering skills for painting; analysis of time-series behavior in EOG signals. Time Stud. **6**, 49–59 (2013) (in Japanese)
8. H. Sawai, K. Tomotake, Y. Ishii, K. Ueno, E. Koyama, A study of evaluating the process of mastering skills including pause and adjustment; psychophysiological and behavioral evaluation using the information of biological signals and subjective scores. in *Proceedings of 2nd IIAI International Conference on Advanced Applied Informatics*, pp. 355–360 (2013)
9. E. Gowen, R.C. Miall, Eye-hand interactions in tracing and drawing tasks. Hum. Mov. Sci. **25**, 568–585 (2006)
10. R.S. Johansson, G. Westling, A. Backstorn, J.R. Flanagan, Eye-hand coordination in object manipulation. J. Neurosci. **21**(17), 6917–6932 (2001)
11. B. Ficher, B. Breitmeyer, Mechanisms of visual attention revealed by saccadic eye movements. Neuropsychology **25**(1A), 73–83 (1987)
12. J. Stern, L. Walrath, R. Goldstein, The endogenous eyeblink. Psychophysiology **21**(1), 22–32 (1984)
13. Y. Ishii, An investigation to quantitatively evaluate psychophysiological state in the process of mastering tasks involving finger dexterous manipulation. The collection of research summaries deptartement of design engineering and management Kyoto institute of technology 2012, pp. 16–17 (2012) (in Japanese)
14. K. Tomotake, Temporal changes characteristics of working results and eye movements in the process of obtaining the skill for the tasks requiring steady movements of the Arm. The collection of research summaries departement of design engineering and management Kyoto institute of technology 2013. pp. 92–95 (2013) (in Japanese)

Frameworks for Adaptive Human Management Systems Based on MDA

Haeng-Kon Kim and Roger Y. Lee

Abstract Along with the boom of Web services and the thriving Model Driven Architecture (MDA), we must consider the growing significance and utility of modeling in the development of software and solutions. The main advantages of MDA are the ability to transform one PIM into several PSMs, one for each platform or technology in which the final system will be deployed, and the automatic code generation that implements the system for those platforms from the corresponding PSMs. Service-oriented architectures (SOA) are also touted as the key to business agility, especially when combined with a model-driven approach. Model-Driven Architecture (MDA) is a well-developed concept that fits well with SOA, but until now it has been a specialized technique that is beyond practical application scope of most enterprises. In this chapter, We describe the initial investigation in the fields of MDA and generative approaches to SOA. Our view is that MDA aims at providing a precise framework for generative software production. Unfortunately many notions are still loosely defined (PIM, PSM, etc.). We propose here an initial exploration of some basic artifacts of the MDA space to SOA. Because all these artifacts may be considered as assets for the organization where the MDA is being deployed with SOA, we are going to talk about MDA and SOA abstract components to apply an e-business applications. We also discuss the key characteristics of the two modeling architectures, focusing on the classification of models that is embodied by each. The flow of modeling activity is discussed in the two architectures together with a discussion of the support for the modeling flows provided by MDA. Our model of framework—a unified modeling architecture—is introduced which illustrates how the two architectures can be

H.-K. Kim (✉)
Department of Computer Engineering, Catholic University of Daegu, Gyeongsan-si, Korea
e-mail: hangkon@cu.ac.kr

R. Y. Lee
Software Engineering and Information Technology Institute, Central Michigan University,
Mt. Pleasant, USA
e-mail: lee1ry@cmich.edu

R. Y. Lee (ed.), *Applied Computing and Information Technology*,
Studies in Computational Intelligence 553, DOI: 10.1007/978-3-319-05717-0_12,
© Springer International Publishing Switzerland 2014

brought together into a synergistic whole, each reinforcing the benefits of the other with case study.

Keywords Model-driven architecture (MDA) · Domain model · Service-oriented architectures (SOA) · Software process improvement · Component based development · Repository

1 Introduction

In order to find an appropriate solution to development and design of those systems an appropriate paradigm seems necessary. The object-oriented and component-based technology has not significantly met the needs of these systems, and may be considered as adding additional complexity to a domain that needs simplification. A new paradigm like service-oriented architecture is necessary. SOA is a paradigm that utilizes services as fundamental elements for developing applications. In order to gain the full benefits of such technology, an effective approach to modeling and designing these complex distributed systems is required. In fact there is not a suitable approach to SOA-based development and little works have been done on this area and most of them are for special applications and specific domains. To exploit the benefits of SOA effectively and duly, we propose an approach that involves MDA into the context [1–4].

Service-oriented architecture (SOA) is an approach to loosely coupled, protocol independent, standards-based distributed computing where software resources available on the network are considered as Services [3]. SOA is believed to become the future enterprise technology solution that promises the agility and flexibility the business users have been looking for by leveraging the integration process through composition of the services spanning multiple enterprises. The software components in a SOA are services based on standard protocols and services in SOA have minimum amount of interdependencies. Communication infrastructure used within an SOA should be designed to be independent of the underlying protocol layer. Offers coarse-grained business services, as opposed to fine-grained software-oriented function calls and uses service granularity to provide effective composition, encapsulation and management of services.

The problems of modeling solutions based on SOA have largely been resolved through the recognition of the importance of loose coupling and the consequent separation of concerns. Service Interfaces are shared amongst models showing the implementation and re-use of the services. Whilst the use of modeling within SOA is well established, it has suffered from the same issues as modeling in other architectures. The abstraction gap between the level of detail expressed in the model and the level of detail expressed in the code is a key issue. Yet it is the abstraction gap which is one of the key targets for the Model Driven Architecture. It seems likely, then, that if SOA and MDA can work together they will add value synergistically, leading to greater benefits than either architecture provides in isolation. Yet the two architectures are distant in terms of the way they address the issues surrounding modeling. SOA focuses on the stereotypical roles of models based on separation of concerns. MDA focuses on levels of abstraction, defining the role of models within a process. The question of the compatibility of these two model architectures remains open.

The service-oriented architecture (SOA) approach and the corresponding web service standards such as the Web Service Description Language (WSDL) [5] and the Simple Object Access Protocol (SOAP) [6] are currently adopted in various fields of distributed application development (e.g. enterprise application integration, web application development, inter organizational workflow collaboration). The service-oriented paradigm offers the potential to provide a fine grained virtualization of the available resources to significantly increase the versatility.

Model driven architecture (MDA) [7] has been proposed as an approach to deal with complex software systems by splitting the development process into three separate model layers and automatically transforming models from one layer into the other:

(1) The Platform Independent Model (PIM) layer holds a high level representation of the entire system without committing to any specific operating system, middleware or programming language. The PIM provides a formal definition of an applications functionality without burdening the user with too much detail.

(2) The Platform Specific Model (PSM) layer holds a representation of the software specific to a certain target platform such as J2EE, Corba or in our case the service oriented Grid middleware.

(3) The Code Layer consists of the actual source code and supporting files which can be compiled into a working piece of software. In this layer, every part of the system is completely specified. MDA theory states that a PIM is specified and automatically transformed into a PSM and then into actual code, thus making system design much easier. The trick, of course, lies in the development of generic transformers capable of generating PSM and code layers from the PIM [8]. E-business application on SOA is a relatively young field of distributed computing and is currently lacking any form of tool support for a model driven approach to software development. This is unfortunate since we believe that due to its high complexity and the high rate of churn in the software technology market, a MDA approach is vital to the adoption of this new technology as in Fig. 1. Only if business logic (i.e. application functionality) developers can more or less effortlessly integrate a new middleware into their system, will a widespread adoption be possible. Furthermore, the developers responsible for the integration of the middleware into the overall system should be able to concentrate on middleware concerns and not have to cope with the business logic as well. This separation of concerns can be greatly facilitated by an appropriate MDA approach.

In this chapter, we present a model-driven approach to SOA modeling and designing complex distributed systems based on MDA. MDA separates the Platform Independent Model (PIM) from the Platform Specified Model (PSM) of the system and transforming between these models is achieved via appropriate tools. The chapter proposes a new approach to modeling and designing service-oriented architecture.

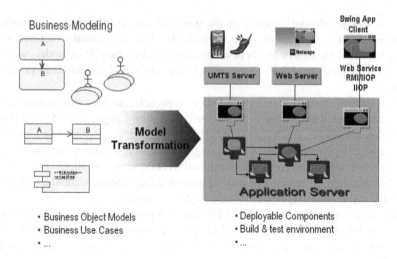

Fig. 1 MDA approaches

In this approach the PIM of the system is created and then the PSM based on SOA is generated (this PSM is a PIM for next level). Then the final PSM based on a target platform (such as Web Services, Jini and so on) is generated. These models are generated with transformation tools in MDA and an approach to the model driven development for e-business applications on SOA is presented. The goal of the approach is to minimize the necessary human interaction required to transform a PIM into a PSM and a PSM into code for a SOA. To further separate the architectures specific components of the PSM from the business specific components of the PSM, a UML e-business Profile is introduced and a separation of the PSM layer into two parts is proposed which make the automated transformations from PIM to PSM to code easier to implement and more transparent for system designers, developers, and users. The separation of concerns introduced on the PSM layer is mirrored on the code layer by the use of Java annotations, allowing the same business code to run in different domains simply by exchanging the annotations and thus decoupling application code and SOA middleware.

2 Related Works

2.1 Modeling Web Services Metadata Based on MDA

Web services are emerging as the perfect framework for application-to-application integration or collaboration, to make these applications available as Web services. To standardize the use of Web services, the World Wide Web Consortium (W3C) proposed the Web Service Description Lan (WSDL) standard, an XML-based language that describes Web service functionality. Essentially, a WSDL file is a language-independent XML-based version of an IDL (Interface Definition Language) file that describes the operations offered by a Web service, as well as the parameters that these operations accept and return. Thus, WSDL has become the standard that supports the description of Web services: What they do, how they should be used, and where they are localized [8, 9].

The WS-Policy framework consists of two specifications: WS-Policy and WS-Policy Attachment.

- The **WS-Policy** specification describes the syntax for expressing policy alternatives and for composing them as combinations of domain assertions. The WS-Policy specification also describes the basic mechanisms for merging multiple policies that apply to a common subject and the intersection of policies to determine compatibility.
- The **WS-Policy** Attachment specification describes how to associate policies with a particular subject. It gives normative descriptions of how this applies in the context of WSDL and UDDI, (Universal Description, Discovery, and Integration), and it provides an extensible mechanism for associating policies with arbitrary subjects through the expression of scopes.

Along with the boom of Web services and the thriving Model Driven Architecture (MDA), we must consider the growing significance and utility of modeling in the development of software and solutions. MDA, which was proposed by the Object Management Group (OMG), is a model-driven framework for software development that proposes to model the business logic with Platform-Independent Models (PIMs) to later transform them on Platform-Specific Models (PSMs) by using transformation guides between the different models. The main advantages of MDA are the ability to transform one PIM into several PSMs, one for each platform or technology in which the final system will be deployed, and the automatic code generation that implements the system for those platforms from the corresponding PSMs.

Because Web services are software components, the development of Web services must exploit the advantages of MDA. To apply the MDA principles in the development of Web services, a modeling process must be considered. According to MDA principles, this modeling activity should result in automatic code generation. If we want to abstract from the platform in which the Web service will be deployed, the code that should be generated is the WSDL document that contains the Web service description in a standard format.

2.2 MDA Main Concepts

The main concepts of the MDA are beginning to be identified [6, 7]. A model represents a particular aspect of a system under construction, under operation or under maintenance. A model is written in the language of one specific meta-model.

A meta-model is an explicit specification of abstraction, based on shared agreement. A meta-model acts as a filter to extract some relevant aspects from a system and to ignore all other details. A meta-meta-model defines a language to write meta-models. There are several possibilities to define a meta-meta-model. Usually the definition is reflexive, i.e. the meta-meta-model is self defined. A meta-meta-model is based at least on three concepts (entity, association, package) and a set of primitive types. The OMG MOF contains all universal features, i.e. all those that are not specific to a particular domain language. Among those features we find all that is necessary to build meta-models and to operate on them. Maintaining a specific tool for the MOF would be costly, so the MOF is aligned on the CORE part of one of its specific metamodels: UML. UML thus plays a privileged role in the MDA architecture. As a consequence, any tool intended to create UML models can easily be adapted to create MOF meta-models. MDA utilizes models and a generalized idea of architecture Platform Independent Model (PIM): describes software behavior that is independent of some platform. Platform Specific Model (PSM): describes software behavior that is specific for some platform. The first step in using MDA is to develop a CIM which describes the concepts for a specific domain. The CIM focuses on the environment and requirements of the system; the details of the structure and processing of the system are hidden or as yet undetermined. The next step involves developing the PIM. The term "platform" can have various meanings and can include one or more system aspects such as operating system, network configurations, and programming language. The meanings of PIM and PSM models are therefore relative to the definition of platform used in the use case. More important than the definition of platform is the recognition that PIMs and PSMs are supposed to separate aspects of program behavior from aspects of implementation. The third step is developing one or more PSMs which characterize a particular deployment of a software application. This could, for example, focus on the properties of a web application, whether the application should be generated in Java or Visual Basic, or whether the installation was for a standalone or networked machine. MDA requires development of explicit transformations that can be used by software tools to convert a more abstract model into a more concrete one. A PIM should be created, and then transformed into one or more PSMs, which then are transformed into code. The mappings between models are meant to be expressed by a series of transformation rules expressed in a formal modeling language. A CIM is a software independent model used to describe a business system. Certain parts of a CIM may be supported by software systems, but the CIM itself remains software independent. Automatic derivation of PIMs from a CIM is not possible, because the choices of what pieces of a CIM are to be supported by a software system are always human. For each system supporting part of a CIM, a PIM needs to be developed first as in Fig. 2.

It is possible for concepts defined in a CIM to be automatically associated with properties defined in a PIM. For example, the concept protein defined in a CIM about proteomics experiments could be associated with PIM concepts such as a help feature that defined protein for users or a drop down list of protein names.

A meta-model in MDA defines a specific domain language. It may be compared to the formal grammar of a programming language. In the case of UML the need to

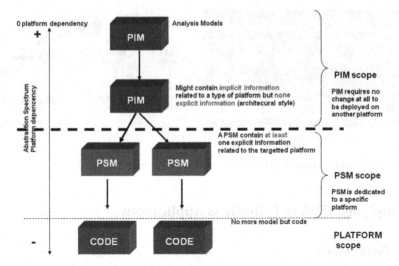

Fig. 2 MDA development process

define variants of the base language was expressed. The UML meta-model was then equipped with extension mechanisms (stereotypes, tagged values, constraints) and this allows defining specialization of the basic meta-models as so called profiles.

The MOF contains features to serialize models and meta-models in order to provide a standard external representation. The XMI standard defines the way serialization is performed. This is a way to exchange models between geographical locations, humans, computers or tools. When a tool reads a XMI serialized model (a UML model for example), it needs to check the version of the meta-model used and also the version of the XMI applied scheme.

2.3 SOA

SOA exposes real dependencies against artificial ones. A real dependency is a state of affairs in which one system depends on the functionality provided by another. Beside real dependencies there are always artificial dependencies in which the system becomes dependent to configurations and various musts other systems expose. The target of SOA is to minimize artificial dependencies (although it can never be completely removed), and maximize real ones. This is done via loosely coupling, and the concept of service. A service is a coarse grain functionality objects, with interfaces expressed via a well defined platform independent language. When using services as computational objects, systems can register, find and invoke each other based on a well defined, every one accepted, language hence no one, highly becomes dependent to another system and a high degree of loosely coupling is achieved.

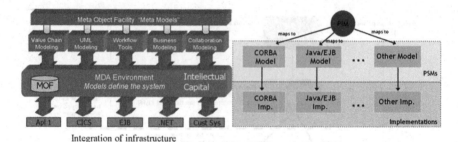

Fig. 3 Architecture for applying MDA to e-business development process

3 Applying MDA to E-Business Applications

3.1 Basic Ideas

The MDA organization may be viewed as a set of artifacts, some being standard building blocks, some being user developed. We may envision, in the not too far future, an organization starting with a hierarchical library of meta-models and extending it as an adaptation to its own local context (models as assets). Model reusability will subsume code reusability, with much more efficiency. This may be seen as orthogonal to code class libraries (e.g. Java, Swing, EJB, etc.). Inside a company, the various business and service models will be developed and maintained to reflect the current situation. Combining a service-oriented modeling architecture with MDA for e-business can bring many unique benefits. First the clear organization of models and information based on the stereotypes derived from the service-oriented architecture and select perspective as development process. Second the productivity, quality and impact analysis benefits of the use of MDA with its emphasis on automation, transformation and synchronization. MS2Web solution for MDA in our approach is uniquely positioned to take advantage of the unified modeling architecture which results from bringing these two key architectures together. MDA combines a uniquely powerful implementation of the web services vision, together with the industry leading solutions for modeling service-based solutions.

Figure 3 shows our architecture for applying MDA to e-business and web service application in this chapter. First it defines the language used for describing object-oriented software artifacts. Second, its kernel is synchronized with the MOF for practical reasons as previously mentioned. There is much less meta-modelers (people building meta-models) than modelers (people building models). As a consequence it is not realistic to build specific workbenches for the first category of people. By making the MOF correspond to a subset of UML, it is possible with some care to use the same tool for both usages. As a consequence the MDA is not only populated by first class MOF meta-models, but also with UML dialects defined by UML profiles for specific purposes languages. This is mainly done for practicality (widening the market of UML tools vendors) and there is some redundancy between UML profiles and MOF meta models (It is even possible to find conversion tools). There are many examples of profiles. Some are standardized by OMG working groups and other are independently defined by user groups or even by individuals. Examples of profiles are "UML for APL 1", " UML for CICS", "UML for Scheduling Performance and Time" (real-time applications), "UML for EJB", "UML testing", "UML for EAI", "UML for QoS and fault tolerance", "UML for Cust Sys".

Figure 3 shows our architecture for applying MDA to e-business and web service application in this chapter. First it defines the language used for describing object-oriented software artifacts. Second, its kernel is synchronized with the MOF for practical reasons as previously mentioned. There is much less meta-modelers (people building meta-models) than modelers (people building models).

As a consequence it is not realistic to build specific workbenches for the first category of people. By making the MOF correspond to a subset of UML, it is possible with some care to use the same tool for both usages. As a consequence the MDA is not only populated by first class MOF meta-models, but also with UML dialects defined by UML profiles for specific purposes languages. This is mainly done for practicality (widening the market of UML tools vendors) and there is some redundancy between UML profiles and MOF meta models (It is even possible to find conversion tools). There are many examples of profiles. Some are standardized by OMG working groups and other are independently defined by user groups or even by individuals. Examples of profiles are "UML for APL 1", " UML for CICS", "UML for Scheduling Performance and Time" (real-time applications), "UML for EJB", "UML testing", "UML for EAI", "UML for QoS and fault tolerance", "UML for Cust Sys".

One important kind of model that is being considered now is the correspondence model. A correspondence model explicitly defines various correspondences that may hold between several models. In the usual case, there are only two models: the source and the target. There may be several correspondences between a couple of elements from source and target. The correspondences are not always between couples of elements and they are strongly typed. There is not yet a global consistent view on correspondence models since this problem is appearing from different perspectives. When the notion of PDM and virtual machine is clarified we may then tackle the definition of a PIM, a model containing no elements associated to a given platform. In other times this was simply called a business model, but as for platform models we need to progress now towards a less naive and a more explicit view. The first idea is that the PIM is not equivalent to a model of the problem. We propose the architectural model for many elements of the solution that may be incorporated in a PIM as long as they don't refer to a specific deployment platform as in Fig. 4.

In our architecture model as in Fig. 3, the PIM of the system is created using UML diagrams by the analyst of the system. The PIM of the system will be designed simply without thinking about services that is pretty simple and is accomplished as Component-Based Development (CBD). The SOA-based PSM (which is a PIM for the next level) would be derived from the present PIM. The way which is used to identify this PSM must be quite different from the one used to identify PSM in component-based systems; because in componentbased systems the patterns which are used to determine the PSM of the system have a specific form. For each

Fig. 4 Architecture for
e-business development in
this chapter

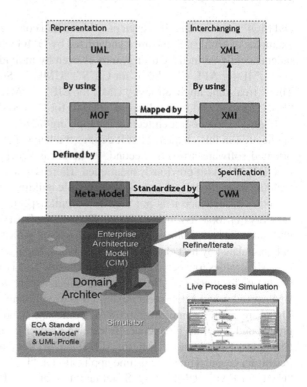

service in e-busimess applications, there is a single instance which manages a set of
resources and consequently, unlike components, services are for the most part state-
less that means need to view a service as a manager object that can create and manage
instances of a type, or set of types. According to above discussion, in our approach
after creating the PIM, this PIM is transformed—with a transformation tool—to
another PIM based on SOA. In this transformation, for each class diagram in PIM
for e-business, a Service Manager is created that manages the Instant Services. This
management involves creation, deletion, updating a service and state management
of services. To complete this transformation, we need some other special patterns for
dealing with associations between classes. When this PIM based on SOA is created,
the PSM of the system can be created based on a target platform such as Web Services, e-business and/or other platforms with transforming tools. Some operations
apply on a single model and are called monadic by opposition to dyadic operations
applying to two models. Operations applying on more than two models are more
rare. Obviously the most apparent components in an MDA workbench are the precise tools composing this workbench. Fortunately in this context we should be able
to propose a rather precise definition of a tool: it is an operational implementation
of a set of operations applicable on specific models. The meta-models supported by
a tool should be exhaustively and explicitly defined.

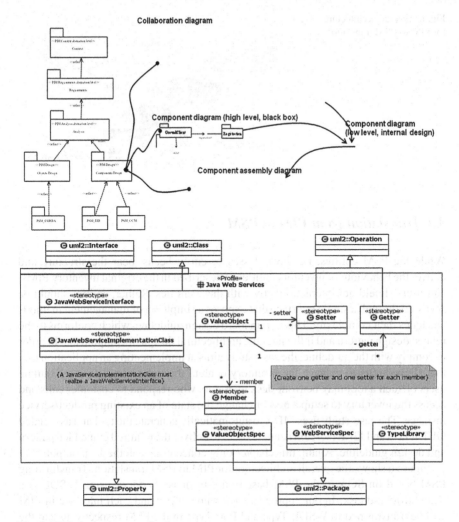

Fig. 5 PIM of e-business applications

3.2 Generating the PIM for E-Business

PIM for e-business application is an abstract design of a computerized solution which does not include any platform specific elements. The core of the platform independent model (PIM) is a UML model? ranging from use cases through classes, interactions, states and other UML elements to the components as in Fig. 5.

Fig. 6 Overall architecture
for PIM to PSM translation

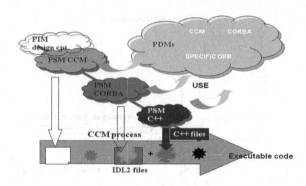

3.3 Translation from PIM to PSM

While the PSM entities, i.e. Java classes or entity beans, bear the structure and
deliver the behaviour of inventory entities as described in the original inventory PIMs,
end-users should not interact directly with these entities. Rather, entities should be
accessed through a single interface that exposes a simple set of management methods
and hides their complexity. This is a standard design guideline, which conforms to the
related design pattern and influences the architectural design of components. In order
to comply with the guideline, the case-study aims at implementing an application tool
that allows users to manage the inventory content through a simple GUI. Example
users of such a tool may be front-desk operators who respond to customer calls and
access the inventory to setup a new or change the state of an existing product/service
instance. The case-study uses MDA to automatically generate the tool and associated
GUI in Java and J2EE (session bean) in order to deliver the required embedded pattern
and design guideline. Again, this chapter only concentrates on the Java outputs.

Figure 6 shows the overall architecture for PIM to PSM translation. Transforming
PSM based on SOA to the PSM based on e-business Services using WSDL is a
straightforward task. In our approach, each value object and each interface in PIM
will be transformed to WSDL Type and Port Type in the PSM respectively and the
parameters of methods will be transformed to the Messages (Input/Output) in the
PSM.

A transformation t transforms a model Ma into another model Mb t: Ma->Mb.
Model Ma is supposed based on meta-model MMa and model Mb is supposed based
on meta-model MMb. We note this situation as: sem(Ma, MMa) sem(Mb, MMb) As
a matter of fact, a transformation is like any other model. So we'll talk about the
transformation model Mt. Mt: Ma->Mb. Obviously since Mt is a model, we postulate
the existence of a generic transformation meta-model MMt, which would be similar
to any other MOF based MDA meta-model:

In some cases the transformation takes some particular form if the source and
target meta-models are in the relation of refinement like a CORBA and a CCM
meta-model. Figure 7 shows the examples of translation interface.

Fig. 7 Example of translation
interface PIM to PSM

```
mapping ParameterToInputPart (in UML2.Parameter) : WSDL.Part {
guard self.direction <> "return" {
     name := self.name;
     element := self.type.resolveByRule(
          "UMLTypeToWSDLElement", WSDL.Element);
     type := self.type.resolveoneByRule(
          "UMLTypeToWSDLType", WSDL.Type);
}

mapping ParameterToOutputPart (in UML2.Parameter) : WSDL.Part {
guard self.direction = "return" {
     name := self.name;
     element := self.type.resolveByRule(
          "UMLTypeToWSDLElement", WSDL.Element);
     type := self.type.resolveoneByRule(
          "UMLTypeToWSDLType", WSDL.Type);
}
```

4 Conclusion

Service Oriented Architecture (SOA) is increasingly important in the business world as b2b transactions become ever more vital to business process out-sourcing and other co-operative activity. The problems of modeling solutions based on SOA have largely been resolved through the recognition of the importance of loose coupling and the consequent separation of concerns. Reinforced by the Supply-Manage-Consume concept, the separate modeling of solutions and services is a well established practice incorporated into advanced development processes that support SOA, including Select Perspective. Service Interfaces are shared amongst models showing the implementation and re-use of the services. Whilst the use of modeling within SOA is well established, it has suffered from the same issues as modeling in other architectures. The abstraction gap between the level of detail expressed in the model and the level of detail expressed in the code is a key issue.

Yet it is the abstraction gap which is one of the key targets for the Model Driven Architecture.

Combining a service-oriented modeling architecture with MDA can bring many unique benefits. First the clear organization of models and information based on the stereotypes derived from the service-oriented architecture and Select Perspective as development process. Second the productivity, quality and impact analysis benefits of the use of MDA with its emphasis on automation, transformation and synchronization. Select Solution for MDA is uniquely positioned to take advantage of the unified modeling architecture which results from bringing these two key architectures together.

In this chapter we introduced an approach to modeling and design of complex distributed systems using SOA and MDA. In fact, to exploit the benefits of SOA effectively and duly, we propose an approach that involves MDA into the context. In this approach the PIM of the system is created and then the PSM based on SOA is generated. Then the final PSM based on a target platform is generated. These models are generated with transformation tools in MDA.

Acknowledgments "This work (Grants No. C0124408) was supported by Business for Cooperative RD between Industry, Academy, and Research Institute funded Korea Small and Medium Business Administration in 2013".

References

1. R. Soley, OMG Staff Strategy Group, Model driven architecture, OMG white paper draft 3.2 (2000), http://www.omg.org/soley/mda.html
2. J.D. Poole, Model driven architecture: vision, standards and emerging technologies, *European Conference on Object-Oriented Programming* (2004), http://www.omg.org/mda/mda_files/Model-Driven_Architecture.pdf
3. M. Rizwan Jameel Qureshi, Reuse and component based development, in *Proceedings of International Conference Software Engineering Research and Practice* (SERP'06), Las Vegas, USA, 26–29 June 2006, pp. 146–150
4. A. Barnawi, M. Rizwan Jameel Qureshi, A. Irshad Khan, A framework for next generation mobile and wireless networks application development using hybrid component based development model. Int. J. Res. Rev. Next Gener. Netw. (IJRRNGN) **1**(2), 51–58 (2011)
5. M. Champion, C. Ferris, E. Newcomer (Iona), David Orchard, Mobile services architecture: W3C working draft (2002), http://www.w3.org/TR/ws-arch/
6. J. Bezivin, S. Hammoudi, D. Lopes, F. Jouault, Applying MDA approach for web service platform, in *Proceedings of the 8th IEEE International Enterprise Distributed Object Computing Conference* (2004)
7. M.N. Huhns, M.P. Singh, Service-oriented computing: key concepts and principles. J. IEEE Internet Comput. **9**(1), 75–81 (2005)
8. A.T. Rahmani, V. Rafe, S. Sedighian, A. Abbaspou, An MDA-based modeling and design of service oriented architecture, in *ICCS*. Part III. LNCS, vol. 3993 (2006), pp. 578–585
9. A. Gokhale, B. Natarajan, Composing and deploying grid middleware web services using model driven architecture. Lect. Notes Comput. Sci. **2519**, 633–649 (2002), http://www.cydex21.com

Author Index

R. Y. Lee (ed.), *Applied Computing and Information Technology*,
Studies in Computational Intelligence 553, DOI: 10.1007/978-3-319-05717-0,
© Springer International Publishing Switzerland 2014

Printed in the United States
By Bookmasters